# SCIENCE WORKSHOP SERIES
## PHYSICAL SCIENCE
# Electricity and Magnetism

## Annotated Teacher's Edition

## Seymour Rosen

**GLOBE BOOK COMPANY**
A Division of Simon & Schuster
Englewood Cliffs, New Jersey

ISBN: 0–8359–0292–7

Printed in the United States of America 1 2 3 4 5 6 7 8 9 0

# CONTENTS

# INTRODUCTION TO THE SERIES

## Overview

The *Science Workshop Series* consists of 12 softbound workbooks that provide a basic secondary-school science program for students achieving below grade level. General competency in the areas of biology, earth science, chemistry, and physical science is stressed. The series is designed so that the books may be used sequentially within or across each of these science areas.

Each workbook consists of approximately 30 lessons. Each lesson opens with a manageable amount of text for students to read. The succeeding pages contain exercises, many of which include photographs or drawings. The illustrations provide students with answers to simple questions. Phonetic spellings and simple definitions for scientific terms are also included to aid in the assimilation of new words.

The question material is varied and plentiful. Exercises such as *Completing Sentences, Matching,* and *True or False* are used to reinforce material covered in the lesson. An open-ended *Reaching Out* question often completes the lesson with a slightly more challenging, yet answerable question.

Easy-to-do laboratory experiments also are included in some lessons. Not isolated, the experiments are part of the development of concepts. They are practical experiments, which require only easily obtainable, inexpensive materials.

Numerous illustrations and photographs play an important role in the development of concepts as well. The functional art enhances students' understanding and relates scientific concepts to students' daily lives.

The workbook format meets the needs of the reluctant student. The student is given a recognizable format, short lessons, and questions that are not overwhelming. The student can handle the stepwise sequence of question material in each lesson with confidence.

The series meets the needs of teachers as well. The workbooks can either be used for an entire class simultaneously, or since the lessons are self-contained, the books can be used on an individual basis for remedial purposes. This works well because the tone of each book is informal; a direct dialogue is established between the book and the student.

## Using the Books

Although each lesson's reading selection may be read as part of a homework assignment, it will prove most effective if it is read during class time. This allows for an introduction to a new topic and a possible discussion. Teacher demonstrations that help to reinforce the ideas stressed in the reading should also be done during class time.

The developmental question material that follows the reading can serve as an ideal follow-up to do in class. The exercises such as *Completing Sentences, True or False* or *Matching* might be assigned for homework or used as a short quiz. The *Reaching Out* question also might be assigned for homework or completed in class to end a topic with a discussion.

## Objectives

The aim of the *Science Workshop Series* is to increase the student's level of competency in two areas: *Science Skills* and *Verbal Skills*. The comprehensive skills matrix on page T-2 highlights the science skills that are used in each lesson.

# SKILLS MATRIX

| Lesson | Identifying | Classifying | Observing | Measuring | Inferring | Interpreting | Predicting | Modeling | Experimenting | Organizing | Analyzing | Understanding Direct and Indirect Relationships | Inductive Reasoning | Deductive Reasoning |
|---|---|---|---|---|---|---|---|---|---|---|---|---|---|---|
| 1 | | | ● | | | ● | | | | ● | | | | ● |
| 2 | ● | | ● | | | ● | | | | ● | | | | ● |
| 3 | ● | | ● | | ● | ● | ● | | | | | | | ● |
| 4 | ● | ● | ● | | | ● | | | | | | | | ● |
| 5 | | | ● | ● | | ● | ● | ● | ● | ● | ● | | | ● |
| 6 | ● | | ● | | ● | ● | ● | | | | | | | |
| 7 | ● | | ● | | | ● | | | | ● | | | | |
| 8 | | ● | ● | | ● | ● | ● | | | ● | | ● | | |
| 9 | ● | ● | ● | | | ● | | | | | | | | |
| 10 | | ● | ● | ● | ● | ● | ● | ● | ● | ● | ● | | ● | ● |
| 11 | | | ● | | | ● | | | | | | | | |
| 12 | ● | | ● | | ● | ● | ● | | | | ● | | | ● |
| 13 | | ● | ● | | | ● | | | | ● | | | | |
| 14 | | | ● | ● | ● | ● | ● | ● | ● | ● | ● | ● | | |
| 15 | ● | ● | ● | | | ● | | | | ● | | | | |
| 16 | ● | | ● | ● | | ● | ● | ● | ● | ● | ● | | ● | ● |
| 17 | ● | | ● | | ● | ● | ● | | | | | | | |
| 18 | | | ● | | ● | ● | ● | | | | ● | | ● | |
| 19 | | | ● | | | ● | | | | | | | | |
| 20 | ● | | ● | | ● | ● | ● | | | | | | | ● |
| 21 | | ● | ● | | | ● | | | | ● | | | | ● |
| 22 | | | ● | | ● | ● | ● | | | | | | | |
| 23 | | | ● | | ● | ● | ● | | | | | | | |
| 24 | | | ● | | ● | ● | | | | ● | ● | | | |
| 25 | | | ● | | ● | ● | ● | | | | ● | | | ● |
| 26 | | | ● | | ● | ● | ● | | | | | | | ● |
| 27 | | | ● | | ● | ● | ● | | | | | | | ● |
| 28 | | ● | ● | | ● | ● | ● | | | ● | | | | |
| 29 | | | ● | | ● | ● | ● | | | | | | | |
| 30 | | ● | ● | ● | | ● | ● | ● | ● | ● | ● | | | ● |
| 31 | | ● | ● | | ● | ● | ● | | | ● | | | | |
| 32 | | | ● | | ● | ● | ● | | | | | | | ● |
| 33 | | | ● | | ● | ● | ● | | | | ● | | | ● |
| 34 | | | ● | ● | | ● | | | | ● | ● | ● | | |
| 35 | | | ● | | | ● | ● | | | | ● | ● | | |
| 36 | | | | ● | ● | ● | | | | | ● | ● | ● | |

## VERBAL SKILLS

An important objective of the *Science Workshop Series* is to give all students—even those with reading difficulties—a certain degree of science literacy. Reading science materials is often more difficult for poor readers because of its special vocabulary. Taking this into account, each new word is introduced, defined, used in context, and repeated in order to increase its familiarity. The development of the term **frequency of vibration** is traced below to illustrate the usage of science vocabulary in the text.

1. The term **frequency of vibration** is introduced in Lesson 3. It is first defined on page 15, then is explained and spelled phonetically on page 16.
2. The measurement of vibration frequency is described on page 17.
3. The *Fill In The Blank* exercise on page 19 requires students to use frequency of vibration in context.
4. The *Matching* exercise on page 20 asks students to recall the units that scientists use to measure frequency.
5. The use of vibration frequency is reinforced in exercises and questions on sound in Lesson 5, pages 29–34.

This stepwise development allows students to gradually increase their working science vocabulary.

Other techniques used to familiarize students with a specialized vocabulary are less formal and allow the student to have fun while reinforcing what has been learned. Several varieties of word games are used. For example:

- A *Crossword Puzzle* appears on page 62.
- A *Word Search* appears on page 106.
- A *Word Scramble* appears on page 155.

## LANGUAGE DIVERSE POPULATIONS

Students with limited English proficiency may encounter difficulties with the core material as well as the language. Teachers of these students need to use ample repetition, simple explanations of key concepts, and many concrete examples from the students' world. Relying on information students already possess helps students gain confidence and establishes a positive learning environment.

To help LEP students with language development, it is important to maintain an open dialogue between the students and the teacher. Encourage student participation. Have students submit written and oral reports. After students read a section of the text, have them explain it in their own words. These strategies will help the teacher be aware of problem areas.

## CONCEPT DEVELOPMENT

In each book the lessons are arranged in such a way as to provide a logical sequence that students can easily follow. Let us trace the development of one concept from the workbook: When light strikes matter, it can be absorbed, reflected, or transmitted.

Lesson 10 familiarizes students with light absorption, reflection, and transmission. Lesson 11 focuses wholly on how light reflection works, and describes the Law of Reflection. Students work with diagrams to gain experience and understanding of this phenomenon. Transmission of light from one medium to another is covered in Lesson 12, on refraction. Again, exercises with diagrams are the means for students to demonstrate their understanding of the concept. Lesson 14 explains absorption of light by opaque objects. As students learn about the colors of objects, they must show that they understand the difference between transmission and absorption.

## SAFETY IN THE SCIENCE LABORATORY

Many of the lessons in the books of the *Science Workshop Series* include easy-to-do laboratory experiments. In order to have a safe laboratory environment, there are certain rules and procedures that must be followed. It is the responsibility of every science teacher to explain safety rules and procedures to students and to make sure that students adhere to these safety rules.

To help students develop an awareness of safety in the science laboratory, a list of Safety Symbols and warnings is included at the end of this workshop.

## USING THE TEACHER'S EDITION

The Teacher's Edition of Physical Science: Electricity and Magnetism, has on-page annotations for all questions, exercises, charts, tables, and so on. It also includes front matter with teaching suggestions for each lesson in the book. Every lesson begins with questions to motivate the lesson. The motivational questions relate to the lesson opener pictures and provide a springboard for discussion of the lesson's science concepts. Following the Motivation are a variety of teaching strategies. Suggestions for *Class Activities, Demonstrations, Extensions, Reinforcements,* and *Cooperative/ Collaborative Learning* opportunities are given.

If a student experiment is included in the lesson, a list of materials needed, safety precautions, and a short explanation of laboratory procedure are given.

The teacher's edition also includes a two-page test, which includes at least one question from each lesson in the book. The test can be photocopied and distributed to students. It begins on the next page. The test's Answer Key is found below.

## ANSWER KEY FOR REVIEW TEST

**Multiple Choice**

1. b    2. d    3. a    4. b    5. b    6. d    7. c    8. d    9. d    10. a

**True or False**

1. false    2. true    3. false    4. false    5. true
6. true    7. false    8. false    9. false    10. true

**Matching**

1. e    2. i    3. f    4. j    5. c    6. h    7. a    8. g    9. d    10. b

**Fill In The Blank**

1. coal, petroleum
2. retina
3. compressed
4. vacuum
5. medium

6. frequency
7. pollution
8. an electric charge
9. a magnetic field
10. heating

### REVIEW TEST

## MULTIPLE CHOICE

*In the space provided, write the letter of the word or words that best complete each statement.*

_____ 1. The pitch of a sound is determined by
   **a)** loudness **b)** vibration speed **c)** speed of sound **d)** bouncing sound

_____ 2. When a beam of light in air strikes water, it can
   **a)** be transmitted **b)** bend **c)** slow down **d)** all of the above

_____ 3. Compared to ultraviolet rays, the frequency of visible light is
   **a)** faster **b)** slower **c)** the same **d)** none of the above

_____ 4. A transparent green filter struck by ordinary visible light will transmit
   **a)** all colors **b)** green **c)** red **d)** blue

_____ 5. The number of electrons flowing past a point in a circuit tells you the
   **a)** wire thickness **b)** current size **c)** pressure or EMF **d)** resistance

_____ 6. A magnet can make a current flow in a wire through
   **a)** resistance **b)** repulsion **c)** attraction **d)** induction

_____ 7. A piece of steel that is hard to make into a magnet will become
   **a)** a permanent magnet **b)** an electromagnet
   **c)** a temporary magnet **d)** a bar magnet

_____ 8. A renewable, nonpolluting energy source for generating electricity is
   **a)** geothermal **b)** nuclear fuel **c)** fossil fuel **d)** solar energy

_____ 9. Light absorbed by a red opaque object includes
   **a)** red and green **b)** all colors **c)** red only **d)** green

_____ 10. In your home, appliances that do not share voltage in the circuit are wired
   **a)** in parallel **b)** along one path **c)** in series **d)** to a battery

## TRUE OR FALSE

*In the space provided, write "true" if the sentence is true. Write "false" if the sentence is false.*

_____ 1. Sound vibrations with lots of energy make soft sounds.

_____ 2. When sound travels to the inner ear, it moves through solid, liquid, and gas.

_____ 3. When light bounces off a surface, it is refracted.

_____ 4. People can either see or hear every part of the electromagnetic spectrum.

_____ 5. Friction can cause static electricity.

_____ 6. Some electrical resistance is found in any metal wire.

_____ 7. Electrons flow in a circuit because they are attracted to the pole of a magnet.

_____ 8. You can make an iron nail magnetic only by touching it with a permanent magnet.

_____ 9. We need transformers to measure the amount of electric current in our homes.

_____ 10. A more efficient home appliance is one that uses less electricity to operate.

## MATCHING

*Match each term in Column A with its description in Column B. Write the correct letter in the space provided.*

| Column A | Column B |
|---|---|
| _____ 1. Lenses | **a)** bounces from a hard surface |
| _____ 2. Retina | **b)** parts of the electromagnetic spectrum |
| _____ 3. Electric current | **c)** strongest at the poles |
| _____ 4. Loads wired in series | **d)** lowers electromotive force |
| _____ 5. Lines of magnetic force | **e)** can make light converge |
| _____ 6. Water power | **f)** flow of electrons |
| _____ 7. Reflected sound | **g)** amplitude of vibration |
| _____ 8. Loudness | **h)** energy source for an electrical generator |
| _____ 9. Step-down transformer | **i)** where an image is formed |
| _____ 10. Microwaves and x-rays | **j)** share voltage in a circuit |

## FILL IN THE BLANK

*Complete each statement using a term from the list below. Write your answers in the spaces provided.*

| | | |
|---|---|---|
| vacuum | compressed | medium |
| heating | coal | an electric charge |
| petroleum | pollution | retina |
| a magnetic field | frequency | |

1. Two fossil fuels are _____ and _____ .

2. The part of the eye where light rays converge is the _____ .

3. As sound travels, vibrating molecules become _____ .

4. Light can travel through a _____ .

5. Sound must travel through a _____ .

6. Energy waves that vibrate fastest have the highest _____ .

7. A problem caused by electric power plants is _____ .

8. Atoms that are not neutral have _____ .

9. Electrons will flow in a conductor that moves through _____ .

10. The sun is a source of energy that can be used for _____ .

# LESSON TEACHING STRATEGIES

## LESSON 1
### What is sound? (pp. 1–6)

**Motivation**   Refer students to the lesson opener art on page 1 and ask these questions:

1.  Is there a difference between music and noise? What is it?

2.  Which is louder—noise or music?

**Class Activity**   Have the class make a variety of sounds with simple materials such as string, rubber bands, paper, balloons, wood blocks, or other similar items that are available. Ask them to record which materials make the greatest variety of sounds.

**Demonstration**   Obtain a tuning fork from a music teacher or physics lab to demonstrate vibrations. Show that the vibrations produced by striking the tuning fork can cause vibration in air or water (see page 5). Use a ping pong ball or piece of styrofoam suspended on a string, or a shallow tray of water, to show that the vibrating tuning fork can cause movement of another object.

## LESSON 2
### How does sound travel? (pp. 7–14)

**Motivation**   Refer students to the lesson opener art on page 7 and ask these questions:

1.  Can you hear sounds when you are underwater?

2.  Do you always hear a sound the very instant it is made?

**Demonstration**   Borrow a stethoscope to show that sound travels through solid objects as well as through air.

**Demonstration**   Show that sound cannot travel without a medium. Place a ringing bell inside a jar that you can evacuate using a vacuum pump (see page 10).

**Class Activity**   Small groups of students should use coiled springs or slinky toys to simulate the movements of longitudinal waves (see page 13).

**Reinforcement**   Strengthen understanding of compression and rarefaction by arranging the class in a long line; use spacing between students to simulate areas of compression and rarefaction.

## LESSON 3
### What is pitch? (pp. 15–22)

**Motivation**   Refer students to the lesson opener art on page 15 and ask these questions:

1.  Can your voice make as many different sounds as this musical instrument can make?

2.  Which of the pictures shows an object making a very low sound?

**Demonstration**   Use a guitar or a piano to show the pitch changes in a musical scale.

**Class Activity**   Have students work in small groups to produce sounds of different pitch by blowing across the tops of beverage bottles (see page 20). If one group uses bottles with one volume (e.g. 12 fl. oz.) have them vary pitch by adding water to different levels, e.g. half full, two thirds full, etc.

**Cooperative/Collaborative Learning**   Have pairs of students investigate effect of thickness and length on quality of sounds produced by plucking strings. Refer to page 21 for description.

## LESSON 4
### What is an echo? (pp. 23–28)

**Motivation**   Refer students to the lesson opener art on page 23 and ask these questions:

1.  How do submarine captains guide their boats underwater if they can't see where they are going?

2.  Does every sound produce an echo?

**Demonstration**   Simulate the effect of different textures on bouncing sound by dropping super balls on materials such as brick, ceramic tile, wood, carpeting, etc.

**Class activity**   If possible, have students produce their own echoes by taking them to an enclosed space with hard walls and no sound absorbing materials.

**Extension** Students should prepare reports on 1) development and use of SONAR in marine navigation; 2) the use of echolocation by animals such as bats and whales; and 3) use of reflected sound in cameras with auto focus.

## LESSON 5
### What is resonance? (pp. 29–34)

**Motivation** Refer students to the lesson opener art on page 29 and ask these questions:

1. Why do skyscrapers vibrate during an earthquake?

2. Have you ever noticed room furnishings vibrating when music is played very loudly?

**Demonstration** Use a pair of tuning forks at the same frequency to show resonance. Vary the distance between the forks to show the effect of separation on resonance.

**Reinforcement** Emphasize that each different kind of material and substance has a different natural frequency of vibration.

**Class Activity** If possible, arrange to expose the strings on a piano. Ask the class to sing different notes in unison. Each note should be sung loudly and briefly. Then listen for the sound of a piano string vibrating. If resonance has taken place, you will hear a piano sound.

## LESSON 6
### What is loudness? (pp. 35–40)

**Motivation** Refer students to the lesson opener art on page 35 and ask these questions:

1. Have you ever had to cover your ears because sound hurt them? What made the sound?

2. Can you suggest a way to measure loudness of the sounds shown in the picture?

**Demonstration** Strike a tuning fork several times with a different amount of force each time to show students relationship between loudness and the force of vibration.

**Class Activity** Have students work in pairs to examine loudness and the amount of displacement in a plucked string. While one student faces away and listens, a second student should pluck a string several times, each time displacing the string a different distance. The student listening should select the loudest sound produced, without watching.

**Reinforcement** Use an oscilloscope and microphone to show the relationship between wave amplitude and loudness.

## LESSON 7
### How do we hear? (pp. 41–45)

**Motivation** Refer students to the lesson opener art on page 41 and ask these questions:

1. How many different sounds can you hear at one time?

2. Are there any sounds that are either too low or too high for you to hear?

**Demonstration** Obtain a model of the human ear to show the parts of the outer, middle, and inner ears.

**Class Activity** Have students make simulated eardrums by stretching a balloon very tightly over one end of a can with both ends removed. Then point the open end of the can toward a sound source such as a loud drum or air horn. Students should see if their simulations pick up vibrations.

## LESSON 8
### How is light different from sound? (pp. 47–52)

**Motivation** Refer students to the lesson opener art on page 47 and ask these questions:

1. Would explosions from combat between space ships in outer space make sound? Why or why not?

2. Why can we see objects and events in outer space, even though we can't hear them?

**Demonstration** Use cards with small holes in them and a flashlight to show that a straight line path of light can be blocked (see page 51). Place a beeper or other small sound source next to the flashlight to show that sound travels around barriers.

**Class Activity** Equip pairs of students with lengths of rubber hose and small flashlights. Have each pair demonstrate to each other that light travels in straight lines.

**Reinforcement** Have the whole class demonstrate a transverse wave by getting in a circle and performing "The Wave." They do this by waving their arms up and down one at a time in a series around the circle. When one student lowers arms, the next student raises them, and so on. This shows vibration perpendicular to the direction of travel.

## LESSON 9
### Where does light come from? (pp. 53–56)

**Motivation**   Refer students to the lesson opener art on page 53 and ask these questions:

1. Does the moon produce its own light? If not, where does moonlight actually begin?

2. Why are some lights very bright, while others are dim?

**Demonstration**   In a darkened room, illuminate a screen with a projector lamp. Ask the students to suggest methods to prove which is the luminous object and which is the illuminated object.

**Cooperative/Collaborative Learning**   Have small groups of students make lists of luminous and illuminated objects in the home and the classroom.

## LESSON 10
### What happens to light when it strikes an object? (pp. 57–62)

**Motivation**   Refer students to the lesson opener art on page 57 and ask these questions:

1. Can you explain why you can see into some windows but not others?

2. What is the difference between a reflection and a shadow?

**Demonstration**   Take four thermometers. Wrap the bulbs of three thermometers in the following material: one, black cloth; two, clear plastic food wrap; three, aluminum foil, shiny side out. Shine a strong lamp on all four thermometers. After 10 minutes, have students put the temperature readings on the board. Have the students tell you how the materials affected the temperature readings. Use this to discuss transmitted, reflected, and absorbed light.

**Extension**   Have students use lights and cards to make shadows on the walls or a screen. Ask students to figure out how to vary the size of the shadows produced by one opaque object and a light source.

## LESSON 11
### What is reflection? (pp. 63–68)

**Motivation**   Refer students to the lesson opener art on page 63 and ask these questions:

1. What is this child doing to see over the hedge?

2. How does a periscope work?

**Demonstration**   Use a strong light, a mirror, and the chalkboard to show how a mirror reflects light beams.

**Classroom Activity**   Have small groups of students work together to demonstrate regular and diffuse reflection using a mirror, a large white sheet of paper, and a light source. Refer to page 68 for description.

**Extension**   Have students use convex and concave mirrors (such as the surfaces of stainless tablespoons) to investigate the difference in the way these surfaces reflect light. Ask students to draw diagrams of incident and reflected rays that might explain this difference.

## LESSON 12
### What is refraction? (pp. 69–75)

**Motivation**   Refer students to the lesson opener art on page 69 and ask these questions:

1. Why does this person's arm look broken?

2. Why does the path of light change as it moves from air to water?

**Class Activity**   Obtain a set of small prisms. Have students work in teams to use the prisms to separate visible light into its component colors. Students should place the prisms on large clean sheets of paper. They should then trace onto the paper the paths of the incoming light and the paths of the rainbows of color produced by the prisms. They should label each band of color they see. They will use this in the next lesson.

## LESSON 13
### What is the spectrum? (pp. 77–82)

**Motivation**   Refer students to the lesson opener art on page 77 and ask these questions:

1. How does a walkie-talkie or a portable telephone send and receive voices?

2. Do x-rays and visible light have anything in common? What is it?

**Demonstration** If possible, obtain some old x-rays from a radiologist or orthopedic surgeon to show to the class. Also obtain some infrared photographs to show that invisible light can be focused to produce an image.

**Class Activity** Have students take the tracings they made in Lesson 12 and figure out which color was bent the most, which one the least. Students should then look at Figure C on page 81 to determine the relationship between the frequency of the colored light and how much it is bent by a prism.

**Extension** Have students prepare reports on images formed by infrared radiation. Topics for research include: infrared photography, remote sensing from satellites, night vision scopes used by the military, etc.

## LESSON 14
### What gives an object its color? (pp. 83–88)

**Motivation** Refer students to the lesson opener art on page 83 and ask these questions:

1. Why are there so many different colors in plants and animals?

2. What happens if you mix two different colors of paint together?

**Class Activity** Provide the students with colored, transparent filters, and light sources. In a darkened room, have them mix beams of different colors to see what happens when colors combine. They should record their results on the chalkboard.

## LESSON 15
### What is a lens? (pp. 89–94)

**Motivation** Refer students to the lesson opener art on page 89 and ask these questions:

1. Can you take a clear, sharp picture of something very close and something far at the same time?

2. Where do you have to hold a magnifying glass in order to use it?

**Demonstration** Use a convex lens to focus a beam of colored light onto a small screen. Demonstrate a measurement of focal length.

**Class Activity** Provide lenses for small groups of students, along with clean sheets of paper that they can use as small portable screens. Have students try to focus an image of various objects onto the paper. They should use a ruler to measure the distance between object and lens, and between lens and screen each time they focus an image. Discuss the results afterwards.

## LESSON 16
### How do we see? (pp. 95–100)

**Motivation** Refer students to the lesson opener art on page 95 and ask these questions:

1. What do your eyes and a camera have in common?

2. Does your eye have muscles? Can you make them stronger with exercise?

**Demonstration** Obtain a model of the human eye to show students the various parts. Demonstrate the path that light takes to reach the retina.

**Class Activity** Have students work in pairs to examine changes in aperture, or pupil diameter, of the eye. Adjust the light in the room from very bright to very dim. Student pairs should observe each other's eyes to note changes in pupil size.

## LESSON 17
### How do eyeglasses help some people see better? (pp. 101–106)

**Motivation** Refer students to the lesson opener art on page 101 and ask these questions:

1. Have you ever had an eye exam? What was the doctor testing?

2. How would you drive a car if you had trouble focusing on distant objects?

**Demonstration** Invite an optician into class to show how lenses are made.

**Class Activity** Borrow an eyechart from an eye doctor. Have the entire class—eyeglass wearers and non-wearers—find the smallest letters they can easily read from a fixed distance.

## LESSON 18
### What is laser light? (pp. 107–112)

**Motivation** Refer students to the lesson opener art on page 107 and ask these questions:

1. Are all lasers used as weapons?

2. What are some ways lasers improve people's lives?

**Reinforcement** Students may be confused by the meaning of in phase and out of phase. You can demonstrate what happens to waves that are out of phase with a rope. Use a long rope to produce two transverse waves moving toward each other from opposite ends. Have students observe what happens to the wave forms when they meet.

**Extension** Have students prepare reports on uses of lasers in the following areas: surgery, holography, repair of fine art, guidance of "smart" weapons.

## LESSON 19
### What is static electricity? (pp. 113–120)

**Motivation** Refer students to the lesson opener art on page 113 and ask these questions:

1. Have you ever combed your hair and found that it stuck out in all directions?

2. How do you get balloons to stick to your walls and ceiling? Why does this happen?

**Demonstration** Rub a glass rod with cloth or fur to produce an excess of charge; hold the rod next to an electroscope to demonstrate the flow of excess charge and repulsion of like charges. Refer to page 118.

**Class Activity** Have students work in pairs to produce static charge using pieces of fur to rub plastic or rubber combs (see pages 116–17). This activity works best in dry, not humid, air.

## LESSON 20
### What is electric current? (pp. 121–128)

**Motivation** Refer students to the lesson opener art on page 121 and ask these questions:

1. Have you ever looked at the electric meter in your home? What does it measure?

2. How can you tell if an electrical outlet is working?

**Demonstration** Cut open an appliance cord, in cross section, to show the wire conductor and the outer, non-conducting insulation. Explain the difference between conductors and insulators.

**Extension** Have students make a simulated battery using strips of copper and zinc inserted into opposite ends of a lemon. They should fasten wires onto each metal strip, or terminal, then connect the wires to the poles of a galvanometer.

**Reinforcement** Explain that in electric current, electrons flow from negative terminal or electrode to positive.

## LESSON 21
### What is a series circuit? (pp. 129–134)

**Motivation** Refer students to the lesson opener art on page 129 and ask these questions:

1. Could one bad bulb make an entire string of colored lights go out?

2. What would you do if you couldn't turn on anything at home because one light bulb blew out?

**Class Activity** Equip students with batteries, switches, and light bulbs and ask them to assemble a circuit similar to the one shown in Figure A on page 131. Ask them to see what happens when they remove one bulb from the circuit.

## LESSON 22
### What is a parallel circuit? (pp. 135–144)

**Motivation** Refer students to the lesson opener art on page 135 and ask these questions:

1. Can you turn on one light or appliance at home without turning on any others?

2. Why is this type of circuit better than the type you put together in Lesson 21?

**Demonstration** Using a parallel circuit similar to the one shown in Figure C on page 139, remove one bulb and ask students to explain why the other remains lit.

**Cooperative/Collaborative Learning** Working in pairs, each student should draw a diagram of a circuit, then show it to a partner. The partner should identify it as either parallel or series, and then assemble the circuit using the components shown in the diagram. Supply batteries, loads, switches, and wire.

## LESSON 23
### What is electrical resistance? (pp. 145–150)

**Motivation** Refer students to the lesson opener art on page 145 and ask these questions:

1. Why does the electric burner get hot?

2. Can you name another conductor of electricity that gets hot?

**Demonstration** Use an ammeter, a source of current, and several different lengths and thicknesses of wire to show the effects of resistance on amount of current.

**Reinforcement** Emphasize that poor conductors of electricity have more resistance than good conductors, and that the length or thickness of the wire are not the only factors in resistance.

## LESSON 24
### What are amperes, volts, and ohms? (pp. 151–156)

**Motivation** Refer students to the lesson opener art on page 151 and ask these questions:

1. What is the danger of high voltage?

2. Where would you measure more ohms, in the dryer's heating coil, or in its power cord?

**Demonstration** Show the difference in the quantities measured by voltmeters and ammeters by wiring them correctly in a circuit.

## LESSON 25
### What are magnets? (pp. 157–162)

**Motivation** Refer students to the lesson opener art on page 157 and ask these questions:

1. What do you use magnets for at home?

2. Can you name home appliances that have magnets in them?

**Class Activity** Ask students to test a variety of objects to see which of them are magnetic and which are not. Refer to the chart on page 160. Afterward, conduct a discussion about the materials that the students selected on their own to test.

**Reinforcement** Explain that magnetic substances don't always show the behavior of magnets. A magnetic substance is one that can be made into a magnet.

## LESSON 26
### How do magnets behave? (pp. 163–170)

**Motivation** Refer students to the lesson opener art on page 163 and ask these questions:

1. Are both ends of a magnet the same? If not, what is the difference?

2. Where would the pull of the magnet feel the strongest?

**Demonstration** Use magnets to show the difference between the two poles and their effects on magnetic objects.

**Class Activity** Have students make maps of a magnetic field using bar magnets, iron filings, and a thin sheet of paper. Refer to page 166 for the procedure. If an instant camera is available, have students take photos of their maps.

## LESSON 27
### Why are some substances magnetic? (pp. 171–176)

**Motivation** Refer students to the lesson opener art on page 171 and ask these questions:

1. Why don't rocks with iron, nickel, or cobalt act like magnets all the time?

2. Can you destroy a magnet? How?

**Demonstration** Break a bar magnet to show that the pieces still behave like bar magnets with two poles. Use the broken ends to pick up nails or paper clips.

**Demonstration** Use the broken bar magnet to make a map of the magnetic fields using iron filings and a piece of paper. Have students compare this magnetic field to the one they mapped in Lesson 26.

## LESSON 28
### What are temporary and permanent magnets? (pp. 177–182)

**Motivation** Refer students to the lesson opener art on page 177 and ask these questions:

1. Is the earth a permanent or temporary magnet? How do you know?

2. Do all magnets need the same amount of strength to do their jobs?

**Demonstration** Show the two alloys alnico

and Permalloy. Point out their different properties.

**Reinforcement** Emphasize that temporary magnets are usually easy to magnetize, but then lose their magnetism quickly. In contrast, permanent magnets are not easy to magnetize.

## LESSON 29

### How can you make a magnet by induction? (pp. 183–188)

**Motivation** Refer students to the lesson opener art on page 183 and ask these questions:

1. Do you think that nails and bolts can become magnets?

2. If so, do they become temporary or permanent magnets?

**Demonstration** Show students how to pass an object through a magnetic field without actually touching the magnet itself.

**Class Activity** Have students use permanent magnets to show induction of magnetism in iron nails. Refer to pages 186–187.

## LESSON 30

### What is an electromagnet? (pp. 189–194)

**Motivation** Refer students to the lesson opener art on page 189 and ask these questions:

1. Can you name devices in your home that have an electromagnet?

2. Are these electromagnets permanent or temporary?

**Demonstration** Induce magnetism in an iron nail using a battery and a wire (see Figures A–D).

**Class Activity** Have students investigate changing the strength of an electromagnet by varying the number of wire coils around an iron nail. Refer to page 194 for the procedure.

## LESSON 31

### What is a transformer? (pp. 195–202)

**Motivation** Refer students to the lesson opener art on page 195 and ask these questions:

1. What does the transformer in a set of electric toy trains do?

2. Why do we need transformers?

**Demonstration** Take apart a transformer to show the arrangement of primary coils and secondary coils around the core.

**Reinforcement** Have a discussion to emphasize the difference between alternating current and direct current.

**Cooperative/Collaborative Learning** Have pairs of students work together to complete the chart on page 201.

## LESSON 32

### What is an induction coil? (pp. 203–206)

**Motivation** Refer students to the lesson opener art on page 203 and ask these questions:

1. What supplies the electricity needed to run a car?

2. Does the battery provide enough power?

**Demonstration** Show an induction coil from an automobile.

**Class Activity** Make a list of the electrical devices in an automobile that require use of an induction coil to boost the voltage of the auto's battery.

## LESSON 33

### How does an electrical generator work? (pp. 207–210)

**Motivation** Refer students to the lesson opener art on page 207 and ask these questions:

1. Where does the electricity in your home come from?

2. How is this electricity made?

**Demonstration** Show induction of an electric current using a bar magnet, a coil of wire, and a galvanometer (see Figure B on page 209).

**Reinforcement** Have a discussion to emphasize that something has to supply power to run a generator. Explain that in your demonstration your hand movements powered the generator.

**Class Activity** Arrange a class field trip to a power plant.

## LESSON 34

### What can be used to power a generator? (pp. 211–216)

**Motivation** Refer students to the lesson opener art on page 211 and ask these questions:

1. Which source of energy shown in the pictures would you choose for a power plant? Why?

2. Do you know what powers the generators that make your electricity?

**Demonstration** Show the basic design of a commercial generator by spinning a wire loop between the poles of a permanent U-shaped magnet. You might attach a galvanometer to the wire to show the presence of a current.

**Class Activity** Ask students to suggest methods for turning the wire loop. Have them prepare a chart to show how nuclear fuel, fossil fuel, and water power can be used to turn the coils in a generator.

**Extension** Assign students to prepare reports about 1) the types of pollution that result from different kinds of power plants, 2) pollution controls in power plants that burn fossil fuels, and 3) what happens to the steam that many power plants use to run the generators.

**Extension** Have the class make steam powered turbines using pinwheels, gas burners, pyrex flasks, and glass tubing. Place some water in the flask, then close the flask with a rubber stopper to which a glass tube has been attached. Direct the glass tube toward a pinwheel, which should be securely positioned above the open end of the glass tubing. Gently boil the water, and observe what the steam does to the pinwheel.

## LESSON 35

### What are alternate sources of electricity? (pp. 217–222)

**Motivation** Refer students to the lesson opener art on page 217 and ask these questions:

1. What would you choose as a source of power for a generator in your home neighborhood—wind or sunlight? Why?

2. Where does the heat that makes a geyser hot come from?

**Demonstration** Show the class some photovoltaic cells used in devices such as solar powered calculators.

**Class Activity** Arrange a visit to see some actual solar collectors in use on a nearby building

## LESSON 36

### How can people conserve energy? (pp. 223–228)

**Motivation** Refer students to the lesson opener art on page 223 and ask these questions:

1. Do you know how much energy you use up at home each day? each week?

2. How can you cut down on the electricity you use?

**Demonstration** Invite a speaker from the local power utility to speak to the class about methods and materials used in home energy conservation. Ask the speaker to tell the class the cost of electricity per kilowatt hour in the area served by the utility.

**Reinforcement** An explanation of the units watt and kilowatt hour are necessary.

**Class Activity** Have students bring energy guide labels from home appliances and EER labels from room air conditioners to class. Have students calculate the average EER for the air conditioners used as examples. Have them calculate the total annual cost for operating all the appliances used as examples, based on the cost of electricity per kilowatt hour in your area.

# SCIENCE WORKSHOP SERIES
## PHYSICAL SCIENCE
# Electricity and Magnetism

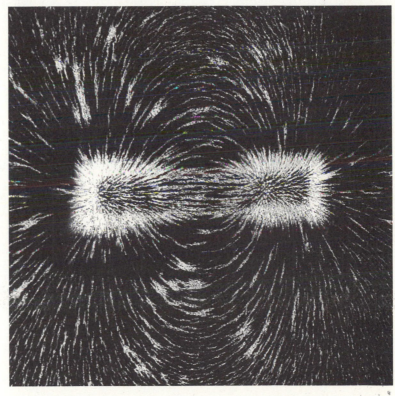

## Seymour Rosen

**GLOBE BOOK COMPANY**
A Division of Simon & Schuster
Englewood Cliffs, New Jersey

## THE AUTHOR

Seymour Rosen received his B.A. and M.S. degrees from Brooklyn College. He taught science in the New York City School System for twenty-seven years. Mr. Rosen was also a contributing participant in a teacher-training program for the development of science curriculum for the New York City Board of Education.

Cover Photograph: Bruce Iverson
Photo Researcher: Rhoda Sidney

Photo Credits:

p.46, Rhoda Sidney Photography
p. 76, Rhoda Sidney Photography
p. 106, Fig. E: Helena Frost
p. 110, Fig. C: Bell Labs
p. 110, Fig. D: Hughes Aircraft Co.
p. 110, Fig. E: Genesco
p. 110, Fig. F: AT&T
p. 111, Fig. G: U.S. Air Force Photo
p. 111, Fig. H: Hughes Aircraft Co.
p. 120, Fig. G: National Oceanographic and Atmospheric Administration
p. 120, Fig. H: New York Public Library
p. 128, Fig. G: U.S. Department of the Interior
p. 144, Fig. G: Bachmann/The Image Works
p. 148, Fig. C: Salt River Project
p. 150: Monkmeyer Press

ISBN:0-8359-0286-2

Printed in the United States of America 1 2 3 4 5 6 7 8 9 10 95 94 93 92 91

Globe Book Company
A Division of Simon & Schuster
Englewood Cliffs, New Jersey

# CONTENTS

# ELECTRICITY

# MAGNETISM

# ENERGY

# Introduction

Imagine trying to go through a day without sound, light, electricity, or magnetism. When you listen to others, look at your surroundings, put on a light, or ride in a motor vehicle, your life is affected by these things.

Sound, light, electricity, and magnetism share important characteristics: they are all forms of energy, and they are all found in nature. People didn't invent them.

In this book, you'll first learn what causes sound and light. Then you'll explore how sound and light behave, how we hear and see, and how people use sound and light in modern life.

You might think that electricity and magnetism are quite different. Perhaps you'll be surprised to discover that they are very closely related. In fact, you'll find out how magnetism is used to make electricity. Along the way, you'll explore how electrical circuits work, and how magnets make the use of everyday machines possible.

Finally, you'll learn how the energy you use every day gets to your home or school. You'll also find out some ways that you can conserve energy, too.

# What is sound?

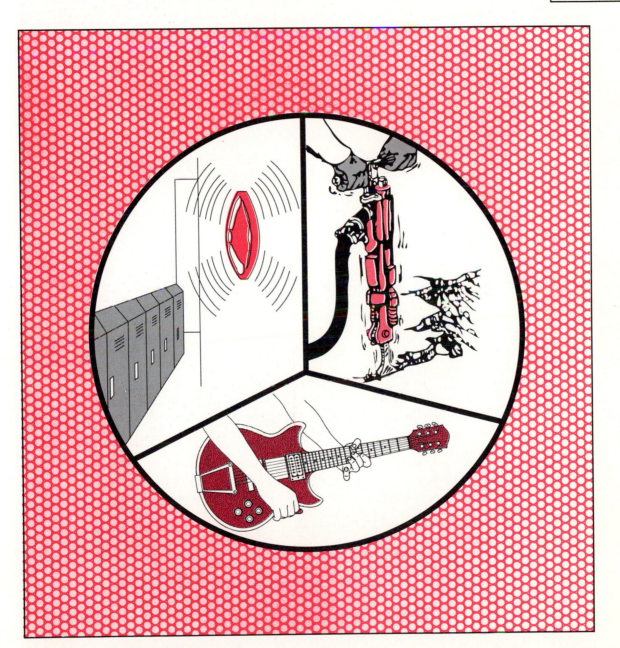

**molecules:**   small parts of matter
**sound:**   a form of energy caused by vibration
**vibrate:**   to move back and forth very rapidly

# LESSON 1 | What is sound?

All learning is done through our senses. We see, we smell. We feel, we taste. We also hear! We hear sounds—all sorts of sounds. We hear words, whistles, squeaks, thumps, music, and many more sounds. Sounds are all around us.

What causes sound? Every **sound** is caused by vibrating matter. To **vibrate** means "to move back and forth very rapidly." A guitar string vibrates when it is plucked. Vibrating air **molecules** make the wind "whistle." Your vocal cords vibrate when you speak. You hit a piece of wood with a hammer, and molecules of both the wood and hammer vibrate.

Now you know what causes sound. But what is sound? Do you remember what energy is? Energy is the ability to make something move. Figures E and F show that sound can make something move. SOUND IS A FORM OF ENERGY.

When energy causes matter to vibrate, molecules of matter move. The sounds you hear are the movements of vibrating matter. Sound vibrations can happen in a gas, a liquid, or a solid.

Sound may be loud, like the horn of a truck, or soft, like the rustle of leaves. Sound may be high, like the chirp of a bird, or low, like thunder. But no matter what kind of sound we are talking about, remember:

• There can be no sound without vibration.

• Whenever there is sound, matter is vibrating.

# MAKING SOUNDS

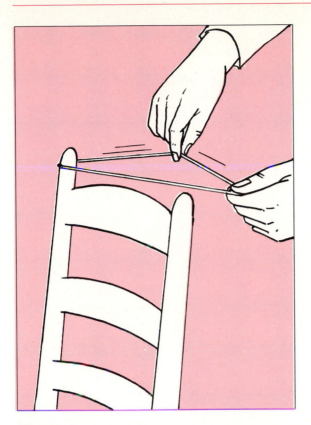

**Figure A**

**Figure B**

Here are simple ways you can make sounds.

Stretch a rubber band around a faucet or chairpost.

Pluck it.

1. Do you hear a sound? _____ yes _____

2. Do you see the rubber band moving?

   _____ yes _____

3. Is it moving slowly or rapidly?

   _____ rapidly _____

4. What do we call this kind of

   movement? _____ vibration _____

5. What is causing the sound here?

   _____ the vibrating rubber band _____

6. What causes any sound?

   _____ vibration _____

Tear off a small piece of paper.

Hold the edge between your lips. Pull it tightly.

Blow hard.

7. Do you hear a sound?

   _____ yes _____

8. What is causing the sound?

   _____ the vibrating paper _____

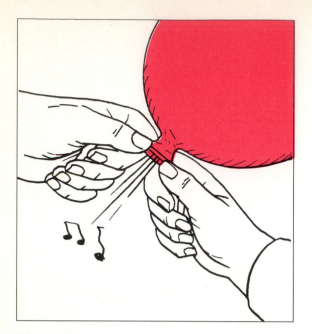

**Figure C**

Blow up a balloon as large as you can. Do not tie it.

Hold the open end tightly. Stretch the neck of the balloon as in Figure C.

Let the air out slowly.

9.  What do you hear? <u>a squeaking sound</u>

10. Two things are vibrating here. What are they? <u>air and the balloon</u>

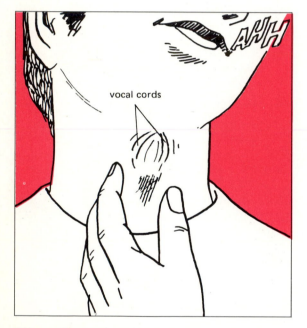

vocal cords

**Figure D**

Place two fingers lightly on your neck as Figure D shows.

Speak.

11. Do you feel vibrations? <u>yes</u>

12. What is vibrating?

    <u>vocal cords</u>

13. Do dogs and cats have vocal cords?

    <u>yes</u>

14. How do you know? <u>they make vocal sounds</u>

*Answer the following.*

1.  Sound is caused by <u>vibration</u>.

2.  There can be no <u>sound</u> without vibration.

3.  If you hear a sound, then you know that something is <u>vibrating</u>.

4

## SOLVE THIS!

Strike a tuning fork with a rubber hammer. The tuning fork is vibrating.

1. How do you know? __you can hear a sound__

Hold the vibrating fork close to a lightweight plastic ball hanging on a string.

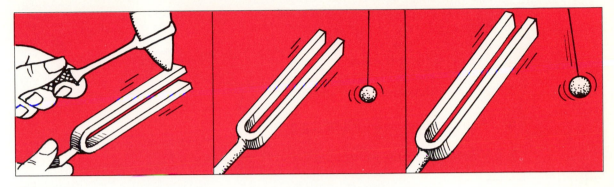

**Figure E**

The tuning fork is not touching the ball. Yet, the ball is moving.

2. How can you explain this? __Tuning fork vibrations cause the air to vibrate. The__ __vibrating air causes the ball to move.__

## WHAT IS HAPPENING?

*Look at Figure F. Write a short, short story explaining what is happening.*

First, the tuning fork is struck. The vibrating fork touches the surface of the water. This causes the water to vibrate. The water vibrations make a splash.

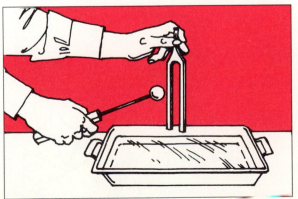 

**Figure F**

## TRUE OR FALSE

*In the space provided, write "true" if the sentence is true. Write "false" if the sentence is false.*

False     **1.** We learn only by seeing.

True     **2.** We learn through all our senses.

True     **3.** Hearing is one of our senses.

False     **4.** We need light to hear.

True     **5.** Hearing depends upon sound.

True     **6.** Sound is caused by vibrating matter.

False     **7.** Vibrating matter moves back and forth slowly.

False     **8.** Our vocal cords are always vibrating.

True     **9.** Vibrations can move from place to place.

False     **10.** Everyone likes the same sounds.

## REACHING OUT

You shake your hand back and forth quickly—and you do not hear a sound. Yet, you can hear the wings of a tiny mosquito buzzing.

1. Why don't you hear the hand movements? The hand is not vibrating fast enough.

2. Why do you hear the mosquito buzzing? The wings are vibrating rapidly enough to make sound.

**Figure G**

**Figure H**

# How does sound travel?

**energy:** the ability to make things move
**medium:** a substance through which sound energy moves

# LESSON 2 | How does sound travel?

A friend calls to you. He is across the street. Yet you hear him clearly. His voice is traveling through the air to your ears.

If you and your friend were on the moon, you could not hear him—even if he shouted. The moon has no air to carry his voice vibrations.

Sound moves from place to place but only where there is matter. Matter is made up of atoms and molecules. Molecules (or atoms) are needed for sound to travel. The vibrations are passed on from molecule to molecule.

A substance through which sound travels is called a **medium** of sound. Solids, liquids, and gases are the mediums of sound.

Sound travels at different speeds through different mediums. The speed depends upon how closely packed the molecules are.

The more closely packed, the faster sound travels.

The more loosely packed, the slower sound travels.

- The molecules of solids are the most closely packed. Sound, therefore, travels fastest through solids.

- The molecules of gases are the most loosely packed. Sound, therefore, travels slowest through gases.

- The molecules of liquids are spaced neither very close nor very far apart. Sound, therefore, travels at an in–between speed through liquids.

The ability to make things move is called **energy.** Sound is a form of energy because it makes matter vibrate.

Sound vibrations move in all directions. The vibrations cause waves. A wave is like a disturbance. Think of a rock being thrown into water. The rock hits the water. The water makes ripples that move outward. Sound waves move in the same way.

# SOUND SPEED IN DIFFERENT MEDIUMS

*Look at Figures A, B, and C. Each stands for a different medium of sound. The dots are the molecules. Study the figures. Answer the questions by figure letters.*

1. The molecules are spaced closest in

    <u>     B     </u> .
    A, B, C

2. The molecules are spaced farthest

    apart in <u>    A    </u> .
            A, B, C

3. The molecules are spaced neither very tightly nor very loosely in

    <u>     C     </u> .
    A, B, C

4. Which is the solid? <u>    B    </u>
                        A, B, C

5. Which is the liquid? <u>    C    </u>
                        A, B, C

6. Which is the gas? <u>    A    </u>
                     A, B, C

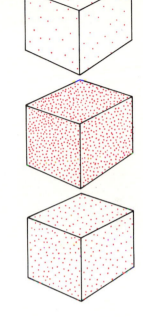

**Figure A**

**Figure B**

**Figure C**

How fast does sound travel through air? It travels about 335 meters (1,100 feet) per second.

7. How long will it take the sound of the airplane to reach the people?

    <u>8 seconds</u>

2,680 meters

**Figure D**

How fast does sound travel through water? Sound travels about 1,500 meters (4,900 feet) per second. In Figure E, the divers on the left are chopping away coral.

8. How long will it take for the chopping sounds to reach the divers on the right?

    <u>½ second</u>

750 meters

**Figure E**

**Figure F**

How fast does sound travel through solids? It depends upon the solid. For example:

- Sound travels through glass at a speed of 3,720 meters (12,200 feet) per second.

- Sound travels through steel at a speed of 5,200 meters (17,060 feet) per second.

9. How far away is the locomotive in Figure F?

    _____15,600_____ meters _____51,180_____ feet

## TEST YOUR UNDERSTANDING

**Figure G**

*Look at Figure H. Then answer the questions.*

There is air in this jar. However, the pump is removing the air.

1. **a)** As the air is removed, the sound

    becomes _____softer_____ .
    softer, louder

    **b)** Why? _Because the medium of sound is being removed._

2. How will you know when just about all the air has been removed? _When you can hardly hear the sound, the air is nearly gone._

3. Sound needs two things: vibrations and a medium. Which of these is being removed here? The _____medium_____ .

10

Complete each statement using a term or terms from the list below. Write your answers in the spaces provided. Some words may be used more than once.

| medium | liquid | gases |
| move | solids | molecules |
| gas | vibrations | solid |
| atoms | directions | |

1. Sound is caused by ____vibrations____ .

2. Sound travels through matter. Any matter through which sound travels is called a

____medium____ of sound.

3. All matter is made up of ____atoms____ and ____molecules____ .

4. Matter is any ____solid____ , ____liquid____ , or ____gas____ .

5. Molecules are most tightly packed in ____solids____ .

6. Molecules are spaced farthest apart in ____gases____ .

7. Sound travels fastest through ____solids____ .

8. Sound travels slowest through ____gases____ .

9. Sound is a form of energy because it can make matter ____move____ .

10. Sound waves move out in all ____directions____ .

## CAN SOUND CHANGE DIRECTION?

Sound vibrations move in all directions. They travel in straight lines. But sound can also turn

corners. How do you know this is true? Because you can hear sounds that come from a

vibrating source around a corner, even when the sounds do not bounce off a wall.

# WHAT ARE SOUND WAVES?

Every vibration produces a sound wave. A sound wave travels through the air in a special way.

You cannot see air molecules. But imagine that you could. What would you see? What does a sound wave do to the molecules? Look at the diagrams and explanations on this page. The dots stand for air molecules.

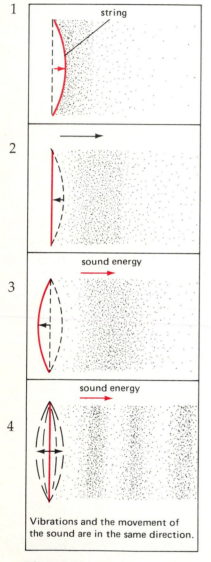

**Figure H**

1. An object, like a guitar string, is plucked. It vibrates.

   First the string moves to one side. It pushes the air molecules in front of it. The molecules crowd together. They become <u>compressed</u> [kum PREST].

2. The compressed molecules move forward. They move away from the vibration that started them moving. The string moves back to its starting position.

3. The string now moves to the opposite side. The molecules become less crowded. They are more spread out. We say the air becomes <u>rarefied</u> [RARE uh fide].

4. The string keeps vibrating back and forth. The air becomes compressed and rarefied over and over again. Soon the air becomes filled with waves of compressed and rarefied molecules. We call these waves <u>longitudinal</u> [lon ji TOOD uhn ul] <u>waves</u>. Sound waves are longitudinal waves. A longitudinal wave vibrates in the same direction as its length.

## SOUND WAVES AND COLLISIONS

When a medium vibrates, the areas of compressed molecules and rarefied molecules change. Molecules in a compressed area collide with molecules in the next rarefied area. As more collisions happen, molecules in a rarefied space become compressed. These collisions between molecules transfer sound energy.

# HOW A LONGITUDINAL WAVE MOVES

A Slinky toy can give you an idea how a longitudinal wave moves.

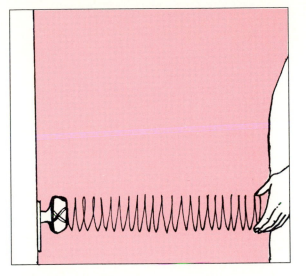

**Figure I**

1. Tie one end of the Slinky to a doorknob. Stretch it out about 1½ meters (5 feet).

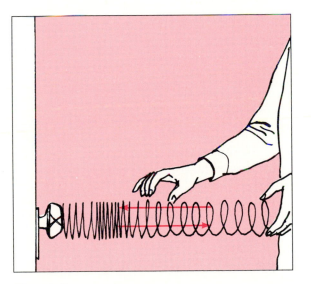

**Figure J**

2. Give your end of the slinky a push with your hand. This compresses—then releases—energy. Notice that the energy moves forward. It moves in the direction of its length, just like a longitudinal wave.

   What happens to the energy when it reaches the end of the Slinky?

   *If you push hard, the energy will travel back.*

   A longitudinal wave acts this way too!

A Slinky shows motion of a longitudinal wave in one direction. Some water in a bucket or plastic basin will show you how waves travel in all directions.

Put several inches of water into a basin or bucket. Wait until the water surface is completely still. Drop a small marble into the center of the body of water. Observe carefully what happens.

What does the marble cause on the water surface? *waves*

In which direction do(es) the wave(s) travel? *outward in all directions*

You cannot see a sound wave. But you can see light. Sound is a form of energy. And energy can change from one form to another.

An instrument called an <u>oscilloscope</u> [uh SILL uh SKOPE] changes sound to electrical energy. Then the electrical energy changes to light energy. The light energy shows the pattern of the sound wave.

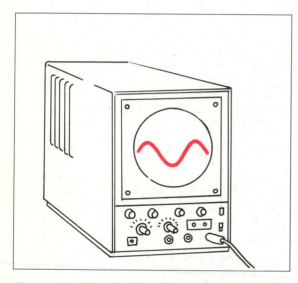

**Figure K**

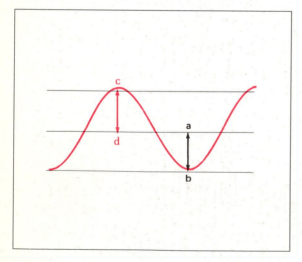

**Figure L**

Every sound has its own special wave pattern. This is the wave pattern of one kind of sound.

Every wave has a <u>compressed</u> part and a <u>rarefied</u> part.

In Figure L, ab stands for the compressed part.

Which part stands for the rarefied part?

*Draw a line where you think it is. Label it cd.*

# What is pitch?

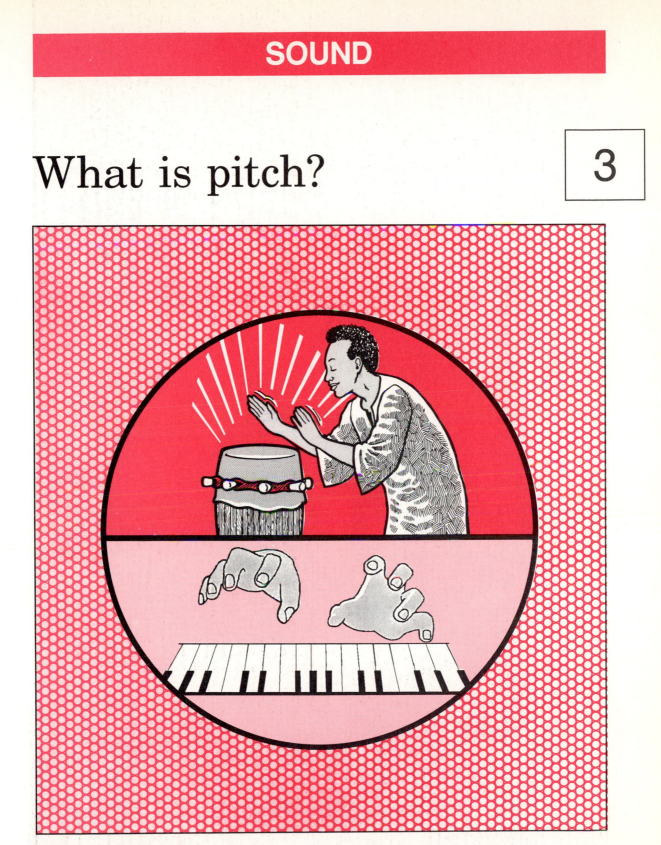

**pitch:** how high or low a sound is
**frequency of vibration:** how often an object vibrates in one second
**hertz:** a unit that measures the frequency of vibration

# LESSON 3 | What is pitch?

The roar of a lion, the squeak of a mouse. The tweet of a flute, the blast of a tuba. A gentle whisper, a deafening shout. Each of these is a different sound. There are many kinds of sounds. Each sound has its own properties. One of the properties of sound is called **pitch.**

WHAT IS PITCH?

The best way to explain pitch is with a musical scale.

You probably know the musical scale. Sing it to yourself. Go ahead, really do it! Do, re, mi, fa, sol, la, ti, do! Notice that the notes become higher and higher.

The musical scale is made up of sounds of different pitches. Each sound has its own pitch. Pitch, then, is how high or low a sound is. Pitch is not how loud or soft a sound is.

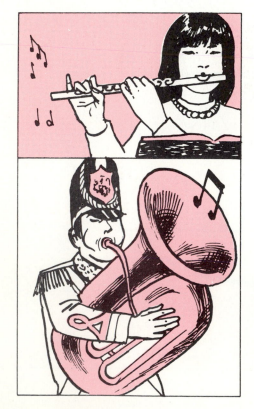

WHAT CAUSES PITCH?

Pitch depends upon how fast an object vibrates in one second. We call this **frequency** [FREE kwen see] **of vibration.**

- The faster or more frequently an object vibrates, the higher is its pitch.

- The slower or less frequently an object vibrates, the lower is its pitch.

A flute sound has a high pitch. A tuba sound has a low pitch.

- A flute sound vibrates faster than a tuba sound.

- A tuba sound has a lower frequency of vibration than a flute sound.

## ABOUT CYCLES AND WAVELENGTH

Frequency of vibration is measured with a unit called **hertz.** One hertz means one vibration per second. One complete vibration is called a cycle. In Figure A, one sound cycle shows up as one wave.

1. How many cycles does Figure A show? _____2½_____

2. A high pitch has _____many_____
   many, few
   cycles every second.

3. A low pitch has _____few_____
   many, few
   cycles every second.

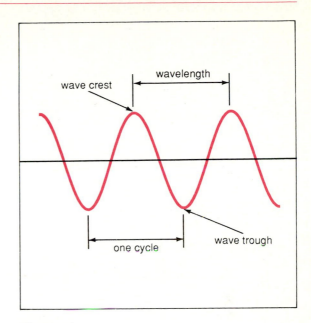

**Figure A**

A wave also has length. The distance between the crests or troughs of two waves that are next to each other is called the wavelength.

*Two wave patterns are shown in Figures B and C. Study the patterns. Answer the questions. Use the figure letters in your answers.*

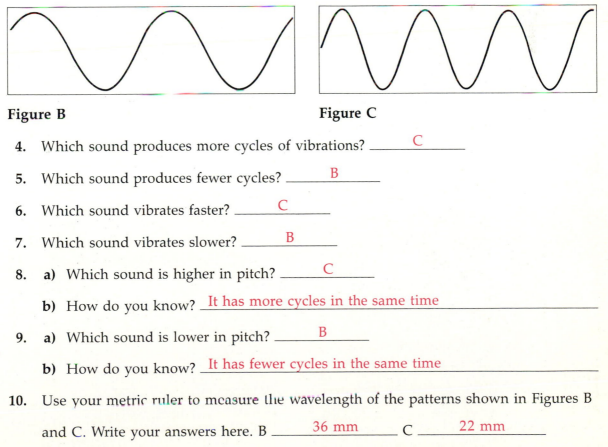

**Figure B**                     **Figure C**

4. Which sound produces more cycles of vibrations? _____C_____

5. Which sound produces fewer cycles? _____B_____

6. Which sound vibrates faster? _____C_____

7. Which sound vibrates slower? _____B_____

8. a) Which sound is higher in pitch? _____C_____

   b) How do you know? __It has more cycles in the same time__

9. a) Which sound is lower in pitch? _____B_____

   b) How do you know? __It has fewer cycles in the same time__

10. Use your metric ruler to measure the wavelength of the patterns shown in Figures B and C. Write your answers here. B _____36 mm_____ C _____22 mm_____

# IDENTIFYING SOUND WAVES

Four sound-wave patterns are shown in Figures D through G. Each one stands for a note on a musical scale.

do  re  mi  fa  sol  la  ti  do

- One pattern stands for the first <u>do</u>.
- One stands for <u>re</u>.
- One stands for <u>mi</u>.
- Still another stands for the last <u>do</u>.

*Match the patterns to the notes. But first answer these questions.*

1. A high-pitched sound vibrates _____<u>faster</u>_____ than a low-pitched sound.
   <br>faster, slower

2. A high-pitched sound has _____<u>more</u>_____ cycles than a low-pitched sound.
   <br>fewer, more

3. A high-pitched sound has a _____<u>shorter</u>_____ wavelength than a low-pitched sound.
   <br>shorter, longer

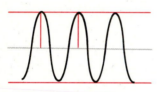

**Figure D**

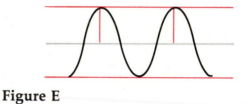

**Figure E**

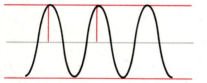

**Figure F**

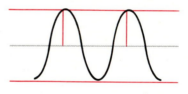

**Figure G**

4. The first <u>do</u> goes with the pattern in Figure _____<u>E</u>_____.
   <br>D, E, F, G

5. <u>Re</u> goes with the pattern in Figure _____<u>D</u>_____.
   <br>D, E, F, G

6. <u>Mi</u> goes with the pattern in Figure _____<u>F</u>_____.
   <br>D, E, F, G

7. The last <u>do</u> goes with the pattern in Figure _____<u>G</u>_____.
   <br>D, E, F, G

8. With your metric ruler, measure the wavelengths of the first <u>do</u> and the last <u>do</u>.

9. The wavelength of the first <u>do</u> measures _____<u>20</u>_____ mm.

10. The wavelength of the last do measures _____10_____ mm.

11. The wavelength of the last do is _____one half_____ the wavelength of the first do.
one half, twice

12. The last do vibrates _____twice as fast_____ as the first do.
twice as fast, half as fast

13. The pitch of the last do is _____twice_____ the pitch of the first do.
half, twice

## FILL IN THE BLANK

*Complete each statement using a term or terms from the list below. Write your answers in the spaces provided. Some words may be used more than once.*

| | | |
|---|---|---|
| soft | lower | loud |
| frequency of vibration | high | low |
| higher | different | fast |

1. Pitch is how _____high_____ or _____low_____ a sound is.

2. Pitch is not how _____loud_____ or _____soft_____ a sound is.

3. Each note of the musical scale has a _____different_____ pitch.

4. Pitch depends upon how _____fast_____ an object vibrates.

5. The faster an object vibrates, the _____higher_____ the pitch; the slower an object

   vibrates, the _____lower_____ the pitch.

6. Another way of saying pitch is _____frequency of vibration_____ .

## MATCHING

*Match each term in Column A with its description in Column B. Write the correct letter in the space provided.*

**Column A**

____d____ 1. cycle

____a____ 2. high-pitched sound

____c____ 3. low-pitched sound

____e____ 4. musical scale

____b____ 5. hertz

**Column B**

a) many cycles

b) a unit that measures frequency of vibration

c) few cycles

d) one complete vibration

e) each note has a different pitch

# UNDERSTANDING PITCH CHANGES

Did you ever blow across the opening of an empty bottle? The moving air made the air in the bottle vibrate. This vibration produced a sound. If you blew across a short bottle, you produced a high pitched sound. If you blew across a tall bottle, you produced a lower pitched sound.

*Figure A shows four columns of air. Look at the figure. Then answer the questions using the letters W, X, Y, Z.*

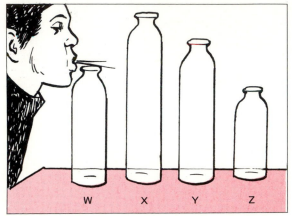

**Figure H**  *Pitch and Columns of Air*

1.  **a)**  Which will produce the highest-

    pitched sound? ___Z___

    **b)**  Why? __It has the shortest air__

    __column__

2.  **a)**  Which will produce the lowest-

    pitched sound? ___X___

    **b)**  Why? __It has the longest air column__

*Underline the correct answer.*

3.  Pitch depends upon

    **a)**  the number of vibrations per second.

    **b)**  the force of the vibrations.

4.  In which column will the air vibrate the fastest? ___Z___

5.  In which column will the air vibrate the slowest? ___X___

# MORE ABOUT PITCH CHANGES

Have you ever plucked a string and made a sound? Try these activities:

## Length and pitch changes

Get a ball of string and cut off two pieces. One should be much longer than the other. Fasten them to a board with eye screws, as shown in Figure I. Pluck each string and listen to the sound. How are the sounds different?

*The sound of the long string is lower in pitch; the short one is higher.*

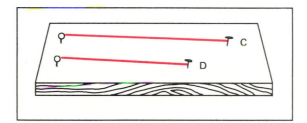

**Figure I**   *In Figure I, strings C and D have the same thickness and the same tightness.*

## Tightness and pitch changes

Pluck one string and listen to the sound. Now turn one eye screw clockwise. This will tighten the string. Pluck again and listen. How does the new sound differ from the first one?

*The new sound is a little higher in pitch than the first one.*

# FILL IN THE BLANK

*Complete each statement using a term or terms from the list below. Write your answers in the spaces provided. Some words may be used more than once.*

|   faster   |   slower   |   higher   |   lower   |
|---|---|---|---|

1.   The longer an object is, the _____*slower*_____ it vibrates.

2.   The shorter an object is, the _____*faster*_____ it vibrates.

3.   The longer an object is, the _____*lower*_____ its pitch.

4.   The shorter an object is, the _____*higher*_____ its pitch.

5.   The tighter an object is, the _____*faster*_____ it vibrates.

6.   The tighter an object is, the _____*higher*_____ its pitch.

## PITCH AND THE PIANO

Figure J shows the strings of a piano. A piano tuner can change the pitch of a key by changing the tightness of a string.

**Figure J**

1. **a)** Which group of strings produces the higher notes? _group H_

   **b)** How do you know? _Because they are shorter than the others_

2. **a)** Which group of strings produces the lower notes? _group G_

   **b)** How do you know? _Because they are longer than the others_

3. When you tighten a piano string, how does the pitch change? _The pitch becomes a little higher._

## HOW VOICE PITCH CHANGES

Voice is produced in the voice box, or larynx [LAR ingks]. The larynx contains the vocal cords. (You have seen the vocal cords in Lesson 1.) The vocal cords can change shape and tightness.

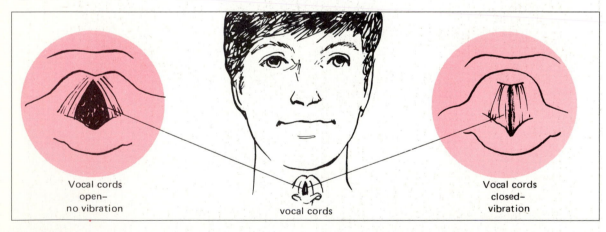

Vocal cords open— no vibration

vocal cords

Vocal cords closed— vibration

**Figure K**

When you are not speaking, the cords are not close together. They do not vibrate.

When you are speaking, the cords are close together. Air passing them from your lungs makes the cords vibrate. The vibrations cause sound.

As you speak, the tightness of the cords changes. The cords also move slightly closer or farther apart. These changes cause changes in pitch.

# What is an echo?

**reflect:** to bounce off
**absorb:** to take in
**echo:** a reflected sound

# LESSON 4 | What is an echo?

You throw a rubber ball against a brick wall or sidewalk and it bounces back. Sound bounces, too. A sound that bounces off a surface is called an **echo.**

Did you ever hear an echo? You can produce one easily. Just stand at least 17 meters (55 feet) from a brick wall. Then clap your hands. Almost immediately you hear the echo of the clap. You can do the same with your voice. Cup your hands and shout in the direction of the wall. You will hear your voice "talking back" to you.

Sound, like a ball, does not bounce off all surfaces.

A ball will bounce easily from a hard surface, like brick, concrete, or stone. But a ball hardly bounces off soft surfaces, like loose sand or mud.

Sound acts in the same way. Sound will bounce, or **reflect,** off hard surfaces but not off soft surfaces. Sound that bounces off a hard surface produces an echo. Sound that hits a soft surface is absorbed.

It is important that some places be as quiet as possible. Heavy carpets, curtains, and soundproofing tiles help reduce the amount of sound. They prevent echoes and **absorb** most sound.

# UNDERSTANDING ECHOES

Look at Figures A through D. Then answer the questions with each.

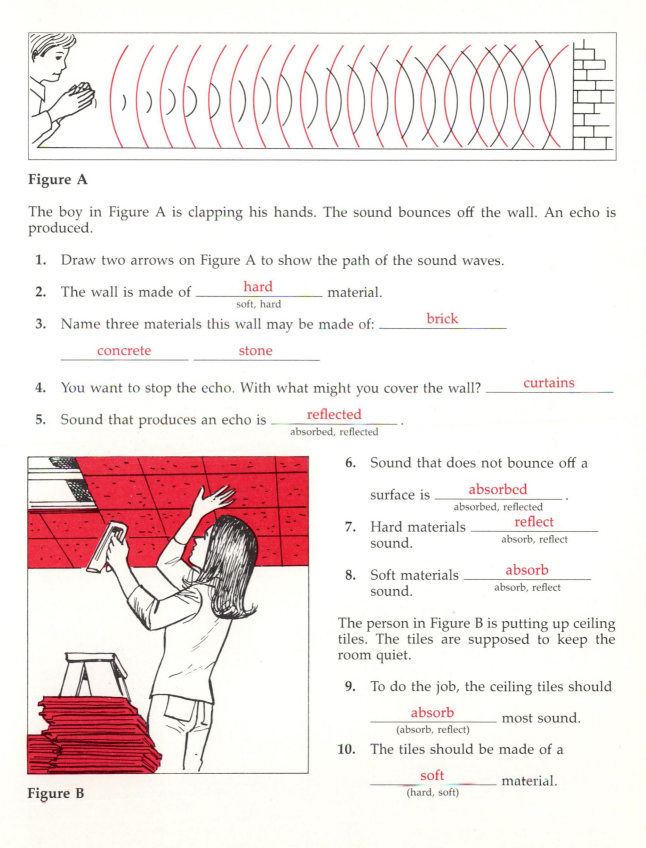

**Figure A**

The boy in Figure A is clapping his hands. The sound bounces off the wall. An echo is produced.

1. Draw two arrows on Figure A to show the path of the sound waves.

2. The wall is made of _____hard_____ material.
   <sub>soft, hard</sub>

3. Name three materials this wall may be made of: _____brick_____

   _____concrete_____   _____stone_____

4. You want to stop the echo. With what might you cover the wall? _____curtains_____

5. Sound that produces an echo is _____reflected_____ .
   <sub>absorbed, reflected</sub>

6. Sound that does not bounce off a

   surface is _____absorbed_____ .
   <sub>absorbed, reflected</sub>

7. Hard materials _____reflect_____ sound.
   <sub>absorb, reflect</sub>

8. Soft materials _____absorb_____ sound.
   <sub>absorb, reflect</sub>

The person in Figure B is putting up ceiling tiles. The tiles are supposed to keep the room quiet.

9. To do the job, the ceiling tiles should

   _____absorb_____ most sound.
   <sub>(absorb, reflect)</sub>

10. The tiles should be made of a

    _____soft_____ material.
    <sub>(hard, soft)</sub>

**Figure B**

**Figure C**                                    **Figure D**

*Look at Figures C and D.*

10. **a.** In which room would your voice make an echo? _____Fig. D_____ .

    **b.** Why? _Because there are no soft materials to absorb sounds, only hard surfaces_

    _to reflect sounds._

11. List the materials that absorb most sounds:

    _carpets, curtains, soundproof tiles, soft furniture_

## FILL IN THE BLANK

*Complete each statement using a term or terms from the list below. Write your answers in the spaces provided. Some words may be used more than once.*

| | | |
|---|---|---|
| an echo | hard | fabric |
| concrete | "soak in" | sound |
| foam rubber | stone | soundproofing |
| "bounce back" | absorb | soft |

1. Absorb means _____soak in_____ .

2. Reflect means _____bounce back_____ .

3. _____Sound_____ can be absorbed or reflected.

4. Reflected sound can produce _____an echo_____ .

5. Sound reflects best off _____hard_____ surfaces.

6. Sound is absorbed by _____soft_____ surfaces.

7. Two materials that absorb sound are _____fabric_____ and _____foam rubber_____ .

8. Two materials that reflect sound are _____concrete_____ and _____stone_____ .

9. Materials that reduce noise are called _____soundproofing_____ materials.

10. Soundproofing materials _____absorb_____ sound.
    <br>                    <sub></sub>one word

# ABSORBS OR REFLECTS?

*Seven materials are listed below. Each one either absorbs sound or reflects sound. Which does each do?*

*Write your answer in the proper boxes.*

|  | Material | Absorbs sound or reflects sound? |
|---|---|---|
| **1.** | cork | absorb |
| **2.** | bathroom tile | reflect |
| **3.** | concrete | reflect |
| **4.** | brick | reflect |
| **5.** | foam rubber | absorb |
| **6.** | fabric | absorb |
| **7.** | plastic (like table tops) | reflect |

# SOMETHING INTERESTING

Have you taken any photographs lately? Reflected sound is now used to focus some cameras.

*Read the explanation and look at Figure E to find out how.*

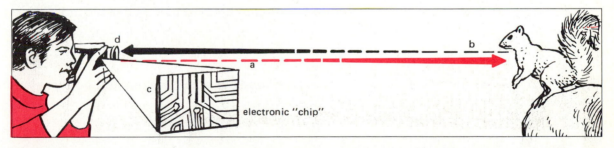

electronic "chip"

**Figure E**

**a.** The camera sends out a very high-pitched sound. You cannot hear the sound—but the camera can "hear" it.

**b.** The sound reflects off a person or object.

**c.** The reflected sound returns to the camera. Electronic parts in the camera "hear" the echo and figure out the distance to the person or object.

**d.** The distance "message" goes to the lens. The lens turns. It focuses the camera perfectly and automatically.

Sound can echo through water as well as air. Echoes in water help ships and submarines find out how deep water is. To find out water depth by using echoes we use a system called <u>sonar</u>.

Sonar also helps locate schools of fish. In time of war, it helps locate enemy submarines.

**Figure F**

This is how sonar works:

1. A ship sends a short-wave sound into the water.

2. When the sound hits the ocean floor, or a school of fish, or a submarine, the sound bounces back to the ship. The ship that sent the original sound gets back the echo.

3. An instrument aboard the ship measures how long the sound takes to make a round trip.

    This time is used to find the depth. How? You figure it out. It's really simple but, slightly tricky. So—don't rush. Think carefully.

**What You Need To Know**   In water, sound travels about 1,500 meters per second.

*Now solve these.*

1. How deep is the water if the sound takes 2 seconds to reach bottom and bounce

    back? _____1500_____ meters

2. How deep for these time measurements?

    a) 1 second? _____750 m_____          d) 8 seconds? _____6000 m_____

    b) 10 seconds? _____7500 m_____       e) 3 seconds? _____2250 m_____

    c) 4 seconds? _____3000 m_____

# What is resonance?

**natural frequency:**   the frequency at which an object vibrates best
**resonance:**   the ability of an object to pick up energy waves of its own natural
frequency

# LESSON 5 | What is resonance?

Did this ever happen to you? You are listening to your favorite music station. Your room "is alive with the sound of music." A certain note is hit—and some object in your room vibrates. It doesn't happen with every note, just a certain note.

Why does that happen? It can be explained in the following way.

Every object has its own frequency of vibration. This is called its **natural frequency.** For example, the natural frequency of one object may be 300 hertz—or 300 vibrations per second. For another object, it may be 325 hertz.

A sound of a certain frequency will cause an object whose natural frequency is the same to vibrate. The object picks up the vibration energy. It vibrates "in sympathy" with the sound. We say that both vibrations are "in tune" with each other.

The ability of an object to pick up energy of its own natural frequency is called **resonance** [REZ uh nance].

Resonance can also be annoying, especially when it causes unwanted sound vibrations. Resonance can be destructive, too. It has been known to crack windows and other glass objects. Some resonance frequencies can make you feel sick.

## UNDERSTANDING RESONANCE

*Study Figure A. Answer the questions.*

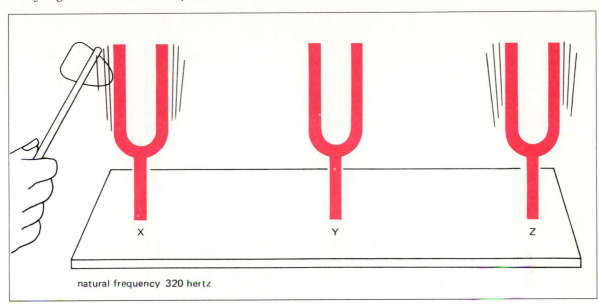

**Figure A**  *Tuning fork X vibrates at 320 hertz*

1.  Which tuning fork is being struck? _____X_____

2.  What is its natural frequency of vibration? __320 hertz__

3.  **a)** Which of the other forks has the same natural frequency? ____Z____

    **b)** How do you know? __It is vibrating in sympathy with X_____

4.  **a)** Which fork does not have the same natural frequency? ____Y____

    **b)** How do you know? __It is not vibrating_____

5.  Which fork would you strike to make fork X vibrate? ____Z____

6.  **a)** Will striking fork Y make fork X vibrate? ____No____

    **b)** Why? __Because they have different natural frequencies and are not in tune.__

7.  **a)** Will striking fork Z make fork X vibrate? ____Yes____

    **b)** Why? __Because they have the same natural frequency__

8.  **a)** Will striking fork Z make fork Y vibrate? ____No____

    **b)** Why? __They do not have the same natural frequency__

9.  Which forks are in tune with each other? __X and Z__

10. An object will pick up energy of its own frequency. What is this called?

    __Resonance__

**Figure B**

Try to get to a piano. If possible, open it so that you can see the strings.

- Press down the pedal on the far right. This frees the strings so they can vibrate freely.

- Sing a single note—like a-a-h-h—but <u>loudly</u>. The string that matches the frequency of that note will vibrate. It will "sing back" to you.

*For question 1, underline the correct answer.*

1.  How can you make a different string vibrate?

    **a)** Change your loudness.

    **b)** <u>Change your pitch.</u>

2.  A piano has 88 keys. How many pitches does a piano have? _____88_____

3.  How many natural frequencies does a piano have? _____88_____

## FILL IN THE BLANK

Complete each statement using a term or terms from the list below. Write your answers in the spaces provided. Some words may be used more than once.

lower  
sympathy  
pitch  
vibrate  

vibrations  
resonance  
hertz  

same  
higher  
frequency  

1. Every sound is caused by _____vibrations_____.

2. The number of times a second an object vibrates is called its _____frequency_____ of vibration.

3. Frequency of vibration determines the _____pitch_____ of a sound.

4. Frequency of vibration is measured in a unit called _____hertz_____.

5. The slower the frequency, the _____lower_____ the pitch. The faster the frequency, the _____higher_____ the pitch.

6. More than one object can have the _____same_____ natural frequency.

7. A sound of a certain natural frequency can cause an object of the same natural frequency to _____vibrate_____.

8. An object will vibrate strongly when it absorbs energy of its own frequency. This is called _____resonance_____.

9. Resonance causes objects to vibrate "in _____sympathy_____" with each other.

## MATCHING

Match each term in Column A with its description in Column B. Write the correct letter in the space provided.

| Column A | | Column B |
|---|---|---|
| __d__ | 1. hertz | a) an object's own vibration speed |
| __a__ | 2. natural frequency | b) no resonance produced |
| __c__ | 3. resonance | c) caused by sympathetic vibration |
| __e__ | 4. 288 hertz sound and 288 hertz object | d) vibrations per second |
| __b__ | 5. 288 hertz sound and 320 hertz object | e) resonance produced |

# TRUE OR FALSE

*In the space provided, write "true" if the sentence is true. Write "false" if the sentence is false.*

__True__ 1. Every pitch has its own frequency.

__True__ 2. A change in frequency changes pitch.

__True__ 3. Every object has its own natural frequency.

__False__ 4. The frequency at which an object vibrates naturally is called its wave height.

__False__ 5. Only one object can have a certain natural frequency of vibration.

__False__ 6. Only a musical string has a natural frequency of vibration.

__True__ 7. A dinner plate has a natural frequency of vibration.

__False__ 8. Objects that vibrate at the same frequency are "out of tune" with one another.

__True__ 9. When a sound reaches an object with the same natural frequency it produces sympathetic vibrations.

__True__ 10. Sympathetic vibrations produce resonance.

# REACHING OUT

How does the distance between a vibrating object and a resonating body affect the loudness of the resonance?

The shorter the distance between a vibrating object and a resonating body, the louder the sound of the resonating body.

# What is loudness?

6

**loudness:**  the amount of energy a sound has
**decibel:**  a unit that measures the loudness of sound

# LESSON 6 | What is loudness?

Sound has pitch. Pitch tells us how high or how low a sound is—like a musical note. Sound also has **loudness.**

There is a wide range of loudness. Some sounds, like a whisper or the chirp of a bird, have a low degree of loudness. They are "soft" sounds. Other sounds, like the roar of a jet or an explosion, have a high degree of loudness. In fact, some sounds are so loud, we have to cover our ears.

During the day you control different degrees of loudness. You change the volume of your radio or TV. You change the loudness of your voice. Sometimes you're told, "Speak up. I can't hear you." Other times it's, "Please lower your voice." Sound familiar? And what about this one? "Please lower your radio. It's blasting my eardrums!"

What causes loudness?

Pitch, you remember, depends upon frequency. The frequency is the number of times a second an object vibrates.

Loudness is different. Loudness depends upon the amount of energy a sound has. The greater the energy, the greater the loudness.

Loudness is measured in **decibels** [DES uh belz]. The higher the decibel number, the louder the sound.

Decibel values start at 0 (zero). A sound of zero decibels is the starting point of human hearing. A sound of 140 decibels may hurt our ears. For example, do you play your radio or stereo very loudly? Listening to very loud music over a long period of time may reduce your hearing—permanently.

## LOUDNESS AND SOUND WAVES

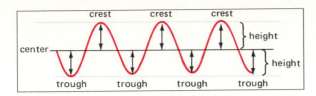

**Figure A**

A sound wave has a high point and a low point. Look at Figure A.

- The high point is called the <u>crest</u>.

- The low point is called the <u>trough</u> [TRAWF].

The distance from the center of a wave to either its high point or low point is called the <u>wave height</u>, or <u>amplitude</u> [am plih TOOD]. Amplitude is a measure of loudness.

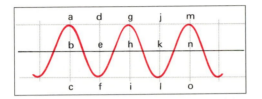

**Figure B**

*Now name the parts of a wave.*

Figure B shows a sound wave. It is labeled with letters a through o. Which part of a sound wave do these letters show? Choose from these terms:

crest          trough
wave height     wavelength

*Write the correct term in the blanks.*

**1.** a     <u>crest</u>

**2.** l     <u>trough</u>

**3.** gh    <u>wave height</u>

**4.** fl    <u>wavelength</u>

**5.** ab   <u>wave height</u>

**6.** kl   <u>wave height</u>

## LOUDNESS AND THE OSCILLOSCOPE

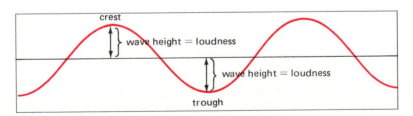

**Figure C** *You can see loudness on an oscilloscope.*

The wave height shows how loud a sound is.

- The <u>higher</u> the wave height is, the <u>louder</u> the sound.

- The <u>lower</u> the wave height is, the <u>softer</u> the sound.

37

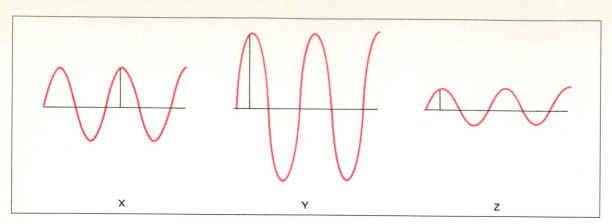

**Figure D**

*Look at the three sets of waves in Figure D. Use your metric ruler to measure the heights of the waves. Now answer these questions.*

1. The wave height of wave X is _____10_____ mm.

2. The wave height of wave Y is _____20_____ mm.

3. The wave height of wave Z is _____5_____ mm.

4. Which wave has the greatest wave height? _____Y_____

5. Which wave has the smallest wave height? _____Z_____

6. Which sound would be the loudest? _____Y_____

7. Which sound would have the highest decibels? _____Y_____

8. Which sound would be the softest? _____Z_____

9. Which sound would have the lowest decibels? _____Z_____

## COMPLETE THE CHART

*The chart lists the average number of decibels that some common sounds have. Four decibel readings have been left out.*

*One wave crest for each of the missing readings is shown in Figures E through H. Read the height of each wave. Then mark down the correct decibel reading on the chart.*

| Sound | Average Number of Decibels |
|---|---|
| whisper | 15 |
| quiet office | 30 |
| classroom | 35 |
| automobile | 45 |
| conversation | 60 |
| light street traffic | 65 |
| heavy street traffic | 75 |
| automobile horn | 100 |
| loud thunder | 110 |

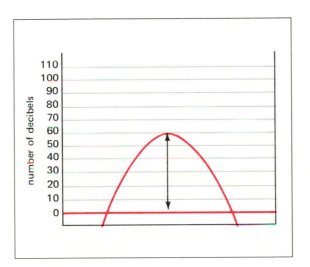

**Figure E**   *Conversation*

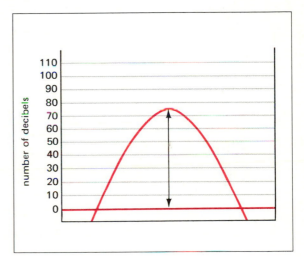

**Figure F**   *Heavy street traffic*

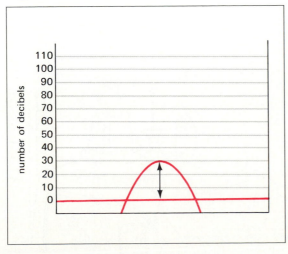

**Figure G**   *Quiet office*

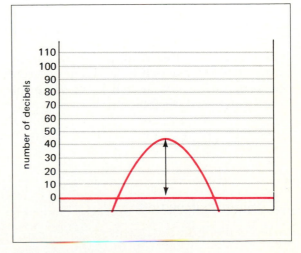

**Figure H**   *Automobile*

## FILL IN THE BLANK

Complete each statement using a term or terms from the list below. Write your answers in the spaces provided. Some words may be used more than once.

| | | |
|---|---|---|
| how rapidly | loud | the amount of energy |
| decibels | less | hertz |
| wave height | loudness | soft |
| more | pitch | |

1. How high or low a sound is (like a note on the musical scale) is called ___pitch___ .

2. Pitch depends upon ___how rapidly___ an object vibrates.

3. Another way of saying "vibrations per second" is ___hertz___ .

4. The property of sound discussed in this lesson is ___loudness___ .

5. Loudness depends upon ___the amount of energy___ a sound has.

6. A loud sound has ___more___ energy than a soft sound.

7. A soft sound has ___less___ energy than a loud sound.

8. Loudness is measured in ___decibels___ .

9. A high-decibel sound is a ___loud___ sound. A low-decibel sound is a ___soft___ sound.

10. The part of a wave that shows loudness is its ___wave height___ .


## REACHING OUT

The sound of a siren changes. Describe the sound after you switch it on. Then describe the sound after you switch it off. (Use the terms pitch and decibels.)

1. Siren switched on. ___The pitch and decibel number increase.___

2. Siren switched off. ___The pitch and decibel number decrease.___

40

**Figure I**

# How do we hear?

7

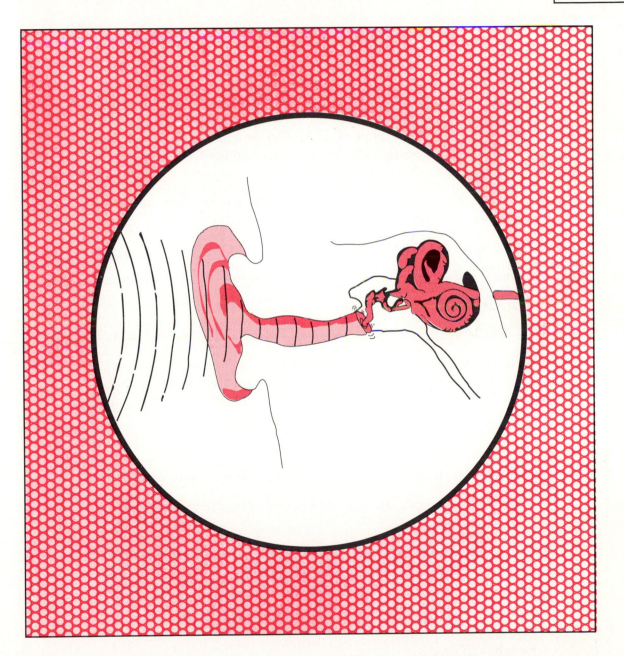

# LESSON 7 | How do we hear?

Everyone knows that we hear with our ears. But the ears don't do it all by themselves. Our ears actually hear the vibrations of sound. But we need the brain to tell us what the vibrations mean. Do vibrations travel to the brain? Not really. The ear changes sound vibrations into nerve signals that go to the brain.

WHAT ARE THE PARTS OF THE EAR? The ear is made up of three sections: the outer ear, the middle ear, and the inner ear.

• The outer ear has two main parts: the auricle [OR i kul] and the ear canal.

The auricle is the part of the ear that sticks out from the side of the head.

The ear canal is a short tube that leads into the head.

• The middle ear contains a thin skin-like tissue called the eardrum.

The middle ear also contains three tiny bones: the malleus [MAL ee us], the incus, and the stapes [STAY peez]. These bones are also called the hammer, anvil, and stirrup.

They are the smallest bones of the body.

• The inner ear contains the cochlea [KOK lee uh] and the semicircular canals. The cochlea is where vibrations are changed into nerve signals.

The semicircular canals are not for hearing. They help us to keep our balance.

HOW DO WE HEAR? Hearing depends upon vibrations passing from one part of the ear to another. Hearing also depends upon the connection from the inner ear to the brain. The exercises on the following pages trace this path.

# HOW SOUND IS HEARD

*Figure A shows the inside of the ear. The explanation traces sound vibrations step by step from outside the ear to the brain. Read the explanations. Find the parts of the ear as you read. Label them on the proper lines.*

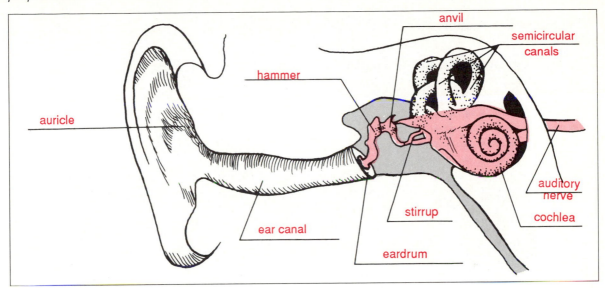

**Figure A**

1. Sound vibrations in the air enter the auricle.

*Label the auricle.*

2. The vibrations then move through the ear canal.

*Label the ear canal.*

3. The vibrations hit against the eardrum. The eardrum vibrates.

*Label the eardrum.*

4. The hammer bone is connected to the inside wall of the eardrum. The vibrating eardrum makes the hammer vibrate.

*Label the hammer.*

5. The hammer passes on the vibrations to the anvil bone.

*Label the anvil.*

6. The anvil passes on the vibrations to the stirrup bone.

*Label the stirrup.*

7. The stirrup passes on the vibrations to the cochlea.

*Label the cochlea. (It is shaped like a snail.)*

8. The cochlea is filled with a liquid. The inside wall is lined with tiny hairs. These hairs extend into the liquid. The vibrations move the liquid and the hairs. This starts a signal in the nerves that are connected to the cochlea. The nerves become a single nerve called the <u>auditory nerve</u>.

*Label the auditory nerve.*

Where does the auditory nerve go? <u>to the brain</u>

9. The semicircular canals are part of the inner ear. But they have nothing to do with hearing.

*Label the semicircular canals.*

What do the semicircular canals do? <u>They help us keep our balance.</u>

## FILL IN THE BLANK

*Complete each statement using a term or terms from the list below. Write your answers in the spaces provided. Some words may be used more than once.*

| | | |
|---|---|---|
| middle ear | auricle | hairs |
| three | ear | outer ear |
| hammer | eardrum | inner ear |
| auditory nerve | ear canal | liquid |
| brain | | |

1. The organ of hearing is the _____ear_____ .

2. Sounds are given meaning in the _____brain_____ .

3. The ear is divided into _____three_____ sections.

4. The sections of the ear are the _____outer ear_____ , the _____middle ear_____ , and the _____inner ear_____ .

5. The part of the outer ear that gathers sound vibrations is called the _____auricle_____ .

6. The tube that leads from the auricle is called the _____ear canal_____ .

7. The part of the ear that vibrates first is called the _____eardrum_____ .

8. The ear bone that vibrates first is called the _____hammer_____ .

9. The cochlea is lined with tiny _____hairs_____ and filled with a _____liquid_____ .

10. The nerve that sends sound messages to the brain is called the _____auditory nerve_____ .

# IDENTIFYING THE PARTS OF THE EAR

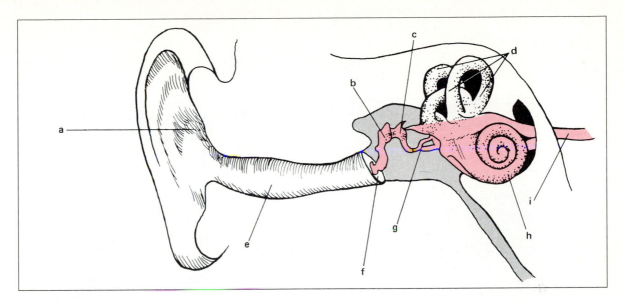

**Figure B**

*Identify the parts of the ear by letter.*

| | | | | | |
|---|---|---|---|---|---|
| __h__ | 1. | cochlea | __d__ | 6. | semicircular canals |
| __e__ | 2. | ear canal | __g__ | 7. | stirrup |
| __c__ | 3. | anvil | __f__ | 8. | eardrum |
| __i__ | 4. | auditory nerve | __b__ | 9. | hammer |
| __a__ | 5. | auricle | | | |

10. Which of these are bones? ___hammer___  ___anvil___

    ___stirrup___

11. List the parts of the outer ear. ___auricle___  ___ear canal___

12. List the parts of the middle ear. ___eardrum___  ___hammer___

    ___anvil___  ___stirrup___

13. List the parts of the inner ear. ___cochlea___  ___semicircular canals___

14. Which part of the inner ear deals with hearing? ___cochlea___

15. Which part of the inner ear deals with balance? ___semicircular canals___

# SCIENCE EXTRA

## Computer Music

Some day, perhaps in 10 years, a rock star will sit down at a computer, write a few bars of music, and ask the computer, "Would you please play this?" The computer will be activated by the composer's voice. Then it will synthesize, or put together, sounds based on the musical symbols written by the composer. Speakers attached to the computer will play those sounds.

How will this work? Computer scientists and music composers are now working to build "smart" sound synthesizers. Linked to computers, these machines may soon be able to make almost any sound you can imagine. This means not just the sounds of musical instruments you know, but also other sounds from nature and around the home. Composers could then combine a wide variety of sounds to make new music that

sounds different from earlier kinds. It may even be possible for new synthesizers to mimic the human voice.

Sound machines linked to computers are already in use in music making. One example is a combination of a grand piano and an electronic sound controller. Because it is made up of two different devices, it is known as a hybrid musical instrument. The piano strings vibrate to produce acoustic sounds. The electronic sound controller is attached to the piano keyboard by wires. With a flip of the switch, the piano keyboard can be used to produce electronic sounds.

Maybe someday you'll go to a rock concert where one performer will sit on stage at a computer and synthesize all the vocal and instrumental music you hear. Awesome!

# How is light different from sound?

**right angle:**  a 90° angle, like any corner of a square
**transverse wave:**  an energy wave that vibrates at right angles to its length
**vacuum:**  the absence of matter

# LESSON 8 | How is light different from sound?

Sound, you have learned, is a form of energy. It has no weight and does not take up space. But sound can do work. It can make things move. That is why sound is a form of energy. Energy is the ability to make things move.

Light is a form of energy, too. But light is different from sound in many ways. How are light and sound different?

1. Sound waves are longitudinal waves. A longitudinal wave vibrates in the same direction as its length. This is what a longitudinal wave looks like.

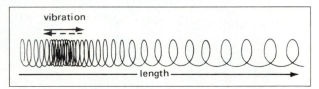

Light waves are **transverse** [trans VURS] **waves.** A transverse wave vibrates at **right angles** to its length.

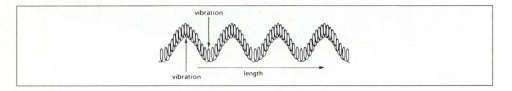

2. Sound waves move only through a medium. That is a solid, liquid, or gas.

Light waves do not need a medium. Light waves can move through a **vacuum.** There is no matter at all in a vacuum, not even air.

3. Sound waves travel through air at about 335 meters (1,100 feet) per second.

Light waves travel much faster. Light travels at a speed of about 300,000 kilometers (186,000 miles) per second. This is the fastest speed in nature. Nothing travels faster than light.

4. Sound waves bend around corners easily. Light waves do not. Light waves travel in straight lines.

## COMPARING SOUND AND LIGHT

Figures A and B show energy waves.

*Look at the figures. Then answer the questions with the figure letter.*

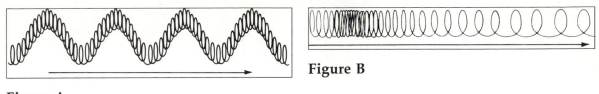

**Figure B**

**Figure A**

1. Which wave vibrates in the same direction as its length? _____B_____

2. Which wave vibrates at right angles to its length? _____A_____

3. Which figure shows a sound wave? _____B_____

4. Which figure shows a light wave? _____A_____

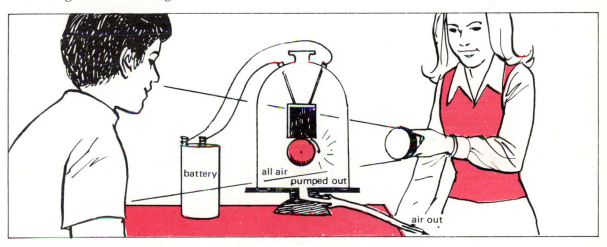

**Figure C**

*Now look at Figure C.*

5. Is there air in the jar? _____No_____

6. What word describes the "absence of air"? _____vacuum_____

7. **a)** Do the students hear the bell ringing? _____No_____

   **b)** Why? _Sound will not travel through a vacuum._____

8. If the flashlight is switched on, the light _____will_____ pass through the jar.
   will, will not

9. Does light pass through a vacuum? _____yes_____

10. Which needs a medium in order to travel, sound or light? _____sound_____

**Figure D**

There is a lightning and thunder storm a distance away.

Lightning and thunder happen at the same time. But you do not experience them at the same time.

11. Which of the following is correct? __b__

   a) You hear thunder before you see the lightning.
   b) You see lightning before you hear the thunder.

12. a) How fast does sound travel through air? ___335 meters per second___

   b) How fast does light travel? ___300,000 kilometers per second___

13. Try to figure out this one.

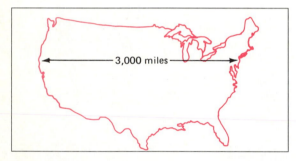

**Figure E**

The United States is about 3,000 miles across.

Light travels about 186,000 miles per second.

How many times can light travel across the United States in just one second?

___62 times___

14. How many round trips could light make across the United States in one second?

___31___

15. *Look at Figure F.*

   a) Will the girl hear her classmate calling? ___yes___

   b) This shows that sound

   ___does___ move around
   <small>does, does not</small>
   corners.

**Figure F**

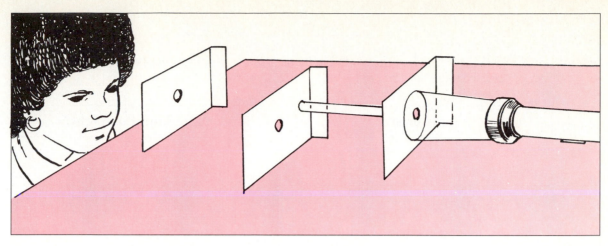

**Figure G**

16. Each piece of cardboard has one hole. Are the holes in a straight line? _____No_____

17. Does the girl see the flashlight bulb? _____No_____

18. This shows that light _____does not_____ move around corners.
    <sub>does, does not</sub>

19. Without moving her head, what can the girl do to see the light? _Move the middle card to the girl's left_

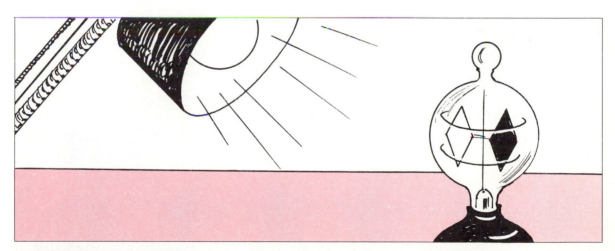

**Figure H**

Figure H shows a <u>radiometer</u>. The light from the bulb is turning the blades.

20. How can you stop the blades from turning? _Turn off the light._

21. What is the definition of energy? _Energy is the ability to make something move._

22. Why are sound and light forms of energy? _They can make things move._

## SOUND OR LIGHT?

*Several characteristics are listed on the chart. Each one belongs to either sound or light. Which one is it? Write either sound or light in the boxes.*

|  | Characteristic | Belongs to Sound or Light? |
|---|---|---|
| 1. | Moves about 335 meters per second | Sound |
| 2. | Transverse waves | Light |
| 3. | Moves around corners | Sound |
| 4. | Moves only in a medium | Sound |
| 5. | Moves about 300,000 kilometers per second | Light |
| 6. | Longitudinal waves | Sound |
| 7. | Does not move around corners; moves in a straight line | Light |
| 8. | Moves in a vacuum | Light |

## REACHING OUT

1. Outer space contains very little matter. It is like a vacuum. How do we know that light can travel through a vacuum? Light reaches us from the sun, moon, and stars.

2. Other forms of energy that you know travel at the speed of light. We use them in our daily lives. What are they?

   Electricity and Microwaves

   Radio and T.V. waves

   Telephone signals

**Figure I**

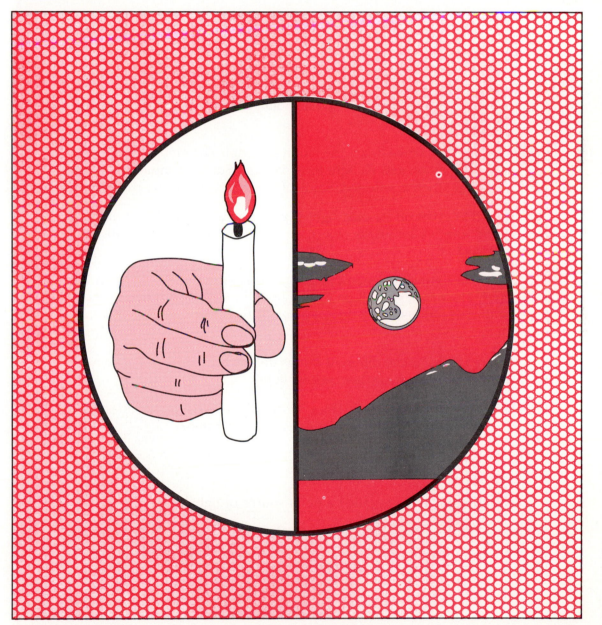

# Where does light come from?

**luminous object:**   an object that gives off its own light
**illuminated object:**   an object that light shines upon

# LESSON 9 | Where does light come from?

Light comes to us from the sun and the moon. But they do not give off light in the same way.

The sun gives off its own light. The moon does not. The moon has no light of its own. The moon gets its light from the sun. Sunlight shines upon the moon. The light then is reflected. Some of the reflected light reaches Earth.

An object that gives off its own light is called a **luminous** [LOO min us] object. The sun is a luminous object. So are switched-on light bulbs and burning wood.

An object that light shines upon is called an **illuminated** [ill OO min AY ted] object. The moon is an illuminated object. In fact, most things you see are illuminated objects. They do not give off their own light. Light shines upon them.

Look around. How many different things do you see? How many give off their own light? How many just receive light?

Look at this book, for example. Does it give off its own light, or does light just shine upon it? Is this book luminous or illuminated?

Some luminous objects are very small. Their light seems to come from a single point. It does not spread out much.

A small luminous body is called a point source of light.

Some luminous objects are large—and close by. Their light comes from many different points. The light spreads out greatly. This kind of luminous body is called an extended light source.

# LUMINOUS AND ILLUMINATED OBJECTS

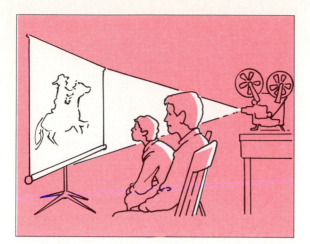

**Figure A**

*Study Figure A. Then answer the questions.*

1. What gives off its own light?
   <u>the projector lamp</u>

2. What receives light from another source? <u>the movie screen</u>

3. Which object is illuminated?
   <u>the screen</u>

4. Which object is luminous? <u>the projector lamp</u>

5. Complete this sentence:

   <u>An illuminated</u> body usually gets its light from
   <small>(A luminous, An illuminated)</small>
   <u>a luminous</u> body.
   <small>(a luminous, an illuminated)</small>

# POINT AND EXTENDED LIGHT SOURCES

*Study Figures B and C. Answer the questions by figure letter.*

One of these diagrams shows a point light source. The other shows an extended light source.

**Figure B**

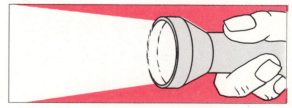

**Figure C**

<u>C</u>   1. Which is the point light source?

<u>B</u>   2. Which is the extended light source?

<u>B</u>   3. Which light spreads out greatly?

<u>C</u>   4. Which light does not spread out much?

<u>C</u>   5. Which light seems to come from one point?

     B       **6.** Which light comes from many points?

     B       **7.** Which would you use to light up a large area?

     C       **8.** Which would you use to light up a small area?

## FILL IN THE BLANK

*Complete each statement using a term or terms from the list below. Write your answers in the spaces provided. Some words may be used more than once.*

     a point              an extended            a luminous
     an illuminated        burns                   illuminated

1. An object that gives off its own light is called _____ a luminous _____ object.

2. An object that receives light is called _____ an illuminated _____ object.

3. A flaming log is an example of _____ a luminous _____ object.

4. You are an example of _____ an illuminated _____ object.

5. Most objects we see are _____ illuminated _____ objects.

6. Any object becomes luminous when it _____ burns _____ .

7. A small luminous object is called _____ a point _____ source of light.

8. A large and close luminous object is called _____ an extended _____ source of light.

9. Light from _____ a point _____ light source does not spread out.

## MATCHING

*Match each term in Column A with its description in Column B. Write the correct letter in the space provided.*

| Column A | | Column B |
|---|---|---|
| e | **1.** light | **a)** an extended light source |
| c | **2.** luminous object | **b)** receives light |
| b | **3.** illuminated object | **c)** gives off its own light |
| d | **4.** a small candle | **d)** a point light source |
| a | **5.** a ceiling light | **e)** a form of energy |

# What happens to light when it strikes an object? 10

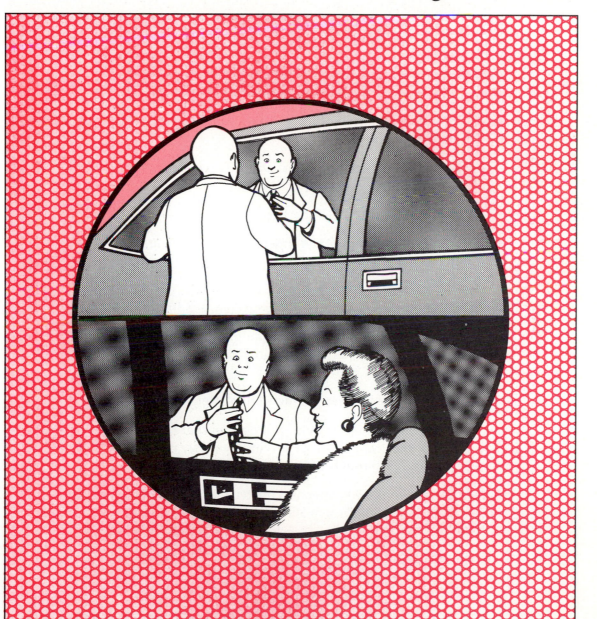

**transmitted:** passed through
**transparent:** letting light and detail pass through; clear
**translucent:** letting light, but no detail, pass through
**opaque:** allowing no light to pass through

# LESSON 10 | What happens to light when it strikes an object?

Light, you have learned, travels in a straight line. It moves along at 300,000 kilometers per second. This is an almost unbelievable speed.

Light also needs no medium in order to travel. Light can move through a vacuum, where there is no matter.

What happens to light when it strikes matter? Three things can happen. The light can be <u>absorbed</u>, <u>reflected</u>, or **transmitted.**

• Light that is absorbed is taken in by the matter it strikes.

Some objects absorb light better than others. Black objects are the best for absorbing light. In fact, black substances absorb all the light that strikes them.

• Light that is reflected "bounces off" the substance it strikes.

A mirror works by reflection. Light strikes an object. The light reflects off the object onto the mirror. The light then reflects off the mirror and into your eyes.

• Light that is transmitted passes through the matter it strikes.

Only certain substances transmit light. Substances that transmit light are said to be **transparent** [tranz PAIR ent]. Window glass, water, and air are transparent. We can see through them clearly.

Some substances—like waxed paper and frosted glass—transmit light. But they also "scatter" the light. We can see light through them but we cannot see any details. Such substances are said to be **translucent** [tranz LOO sent].

Substances like wood and metal do not transmit light. We cannot see through them at all. They are said to be **opaque** [oh PAYK].

# HOW LIGHT STRIKES MATTER

Look at Figures A through F. Each one shows light being absorbed, or reflected, or transmitted. Which does each show? Answer by writing a sentence under each figure. Start the sentence with "Light is being . . ."

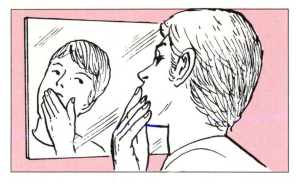

**Figure A**

1. <u>Light is being reflected by the mirror.</u>

**Figure B**

2. <u>Light is being transmitted through water.</u>

**Figure C**

3. <u>Light is being reflected by the water.</u>

**Figure D**

4. <u>Light is being absorbed by the road.</u>

**Figure E**

5. <u>Light is being absorbed by the leaves.</u>

**Figure E**

6. <u>Light is being transmitted through the glass.</u>

*Now answer these questions.*

7. Light that bounces off a substance is _____reflected_____ .

8. Light that passes through a substance is _____transmitted_____ .

9. Light that is taken in by a substance is _____absorbed_____ .

## TRANSPARENT, TRANSLUCENT, OR OPAQUE?

*Look at Figures G, H, and I.*
*Each one shows an object that is either transparent, translucent, or opaque. Which is which? Write the correct term under each figure.*

**Figure G**

1. translucent

**Figure H**

2. opaque

**Figure I**

3. transparent

*Now answer these questions.*

4. Matter that blocks light is _____opaque_____ .

5. Matter that transmits light but no detail of that light is _____translucent_____ .

6. Matter that transmits light along with detail of that light is _____transparent_____ .

## FILL IN THE BLANK

*Complete each statement using a term or terms from the list below. Write your answers in the spaces provided. Some words may be used more than once.*

| | | |
|---|---|---|
| transparent | transmitted | opaque |
| a desk | frosted glass | window glass |
| absorbed | translucent | reflected |

1. Three things can happen to light when it hits matter. It can be _____reflected_____ ,

   or _____transmitted_____ , or _____absorbed_____ .

2. Light that is taken in by matter is _____absorbed_____ .

3. Light that bounces off matter is _____reflected_____ .

4. Light that passes through matter is _____transmitted_____ .

5. A substance that transmits light as well as detail is said to be _____transparent_____ .

6. A substance that blocks light is said to be _____opaque_____ .

7. A substance that transmits light but no detail of that light is said to be

   _____translucent_____ .

8. An example of a transparent object is _____window glass_____ .

9. An example of an opaque object is _____a desk_____ .

10. An example of a translucent object is _____frosted glass_____ .

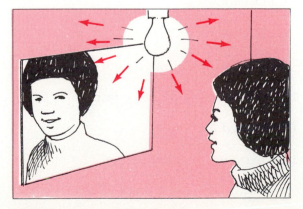

**Figure J**

## REACHING OUT

The girl in Figure J is looking at a mirror. She is seeing a reflection of herself.

Trace the path light takes in order for the girl to see herself. Use lines with arrows.

(Hint: You need three lines.)

# CROSSWORD PUZZLE

*Use the clues to complete the crossword puzzle.*

|   |   |   |   |   |   |   |   |   |   |
|---|---|---|---|---|---|---|---|---|---|
| ¹T | R | ²A | N | ³S | V | E | R | ⁴S | E |   |
| R |   | B |   | P |   |   | E |   | ⁵R |   |
| ⁶A | N | S | W | E | R |   | ⁷W | A | V | E |
| N |   | O |   | E |   |   |   |   | F |   |
| S |   | ⁸R | E | D |   | ⁹H | ¹⁰I | L |   |
| ¹¹M | O | B |   | ¹²O | P | A | Q | U | E |   |
| I |   | ¹³U | ¹⁴P |   | I |   |   | C |   |   |
| ¹⁵T | R | A | N | S | P | A | R | E | N | T |

## Clues

**Across**

1. Light travels as this kind of wave
6. What you give to a question
7. Longitudinal or transverse
8. Color of fire
9. Greeting
11. Crowd
12. Allowing no light to pass through
13. Opposite of down
15. Allowing light and detail to pass through

**Down**

1. An object may absorb, reflect, or _____ light
2. Soak in
3. 300,000 kilometers per second is the _____ of light
4. Ocean
5. Bounce off
9. Natural covering on your head
10. A kind of intelligence test
13. United Nations
14. Extra message in a letter

# What is reflection?

11

**ray:** a single beam of light
**incident ray:** a ray of light that strikes an object
**reflected ray:** a ray of light that is bounced off an object
**normal:** a line that makes a right angle to a surface
**angle of incidence:** the angle between the incident ray and the normal
**angle of reflection:** the angle between the reflected ray and the normal
**Law of Reflection:** the angle of incidence is equal to the angle of reflection

# LESSON 11 | What is reflection?

How does a ball bounce back to you after you throw it against a wall? It depends upon how you throw it. If you throw the ball straight on, it will bounce back straight on. If you throw it at an angle, it will bounce back at an angle.

Light, you know, can bounce. "Bounced" light is reflected light. We can predict how reflected light will behave. Just follow the explanation.

A single beam of light is called a light **ray.** Light is made up of many, many light rays. But let us look at one light ray.

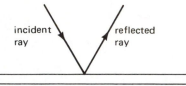

This is a single light ray. It is hitting a flat mirror at an angle. Then it is bouncing off. It is reflecting.

The ray that hits the mirror is called the **incident** [IN si dent] **ray.**

The ray that bounces off the mirror is called the **reflected ray.**

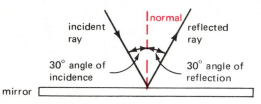

Now let's draw a line that makes a right angle (90 degrees) where the incident ray hits the mirror. This line is called the **normal.**

• The angle between the incident ray and the normal is called the **angle of incidence.**

• The angle between the reflected ray and the normal is called the **angle of reflection.**

The **Law of Reflection** states that "the angle of incidence is equal to the angle of reflection."

In the example on this page, the angle of incidence is 30 degrees. The angle of reflection, then, is also 30 degrees.

# REFLECTING RAYS

Two reflecting rays are shown in Figures A and B. Identify the parts shown by number. Choose from the following:

incident ray
reflected ray

normal
angle of incidence
angle of reflection

*Write your answers next to the correct numbers.*

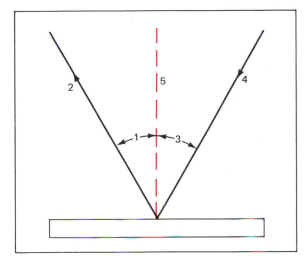

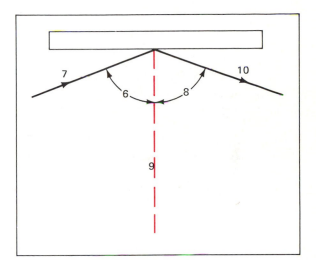

**Figure A**                              **Figure B**

1. _angle of reflection_          6. _angle of incidence_

2. _reflected ray_                7. _incident ray_

3. _angle of incidence_           8. _angle of reflection_

4. _incident ray_                 9. _normal_

5. _normal_                       10. _reflected ray_

11.  State the Law of Reflection. _The angle of incidence is equal to the angle of_ _reflection._

12.  Which of the angles above are equal? (Use numbers.)

  **a)** In Figure A, _1_ and _3_ are equal.

  **b)** In Figure B, _6_ and _8_ are equal.

**Something Extra**

If you have a protractor, measure the angles in Figures A and B. What degrees do the angles

measure? Figure A _30°_

  Figure B _70°_

# KINDS OF REFLECTIONS

There are two kinds of reflections: regular and diffuse [di FYOOS]. What are the differences? Find out for yourself. It's easy! Figures C and D show the two kinds of reflection. They also show light rays all coming from a single source.

*Study each figure. Then answer the questions that go with each.*

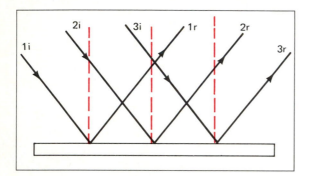

**Figure C**  *Regular reflection*

1.  Figure C shows _____ reflection.
    <br>regular, diffuse: **regular**

2.  A surface that gives a regular reflection is _____ .
    <br>even, uneven: **even**

3.  Every ray has its own normal. In regular reflection, the normals _____ face in the same direction.
    <br>do, do not: **do**

4.  In a regular reflection . . .

    **a)**  every angle of incidence _____ the same.
    <br>is, is not: **is**

    **b)**  every angle of reflection _____ the same.
    <br>is, is not: **is**

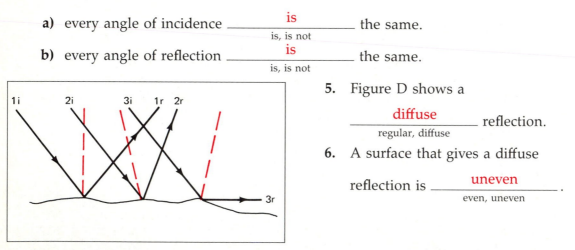

**Figure D**  *Diffuse reflection*

5.  Figure D shows a _____ reflection.
    <br>regular, diffuse: **diffuse**

6.  A surface that gives a diffuse reflection is _____ .
    <br>even, uneven: **uneven**

7.  In a diffuse reflection, the normals _____ face in the same direction.
    <br>do, do not: **do not**

8.  In a diffuse reflection . . .

    **a)**  every angle of incidence _____ the same. (Careful, remember where the angle of incidence is!)
    <br>is, is not: **is not**

    **b)**  every angle of reflection _____ the same.
    <br>is, is not: **is not**

66

## What do you think?

9. Which kind of reflection do you think a mirror gives, regular or diffuse?

   _____regular_____

10. Hold your book up and look at this page.

    a) Does the page reflect like a mirror? _____no_____

    b) This shows that paper gives a _____diffuse_____ reflection.
       <sub>regular, diffuse</sub>

11. Run your hand over this page. To your sense of touch, paper is _____smooth_____.
    <sub>rough, smooth</sub>

12. To light, the surface of the paper is _____uneven_____.
    <sub>even, uneven</sub>

## FILL IN THE BLANK

*Complete each statement using a term or terms from the list below. Write your answers in the spaces provided. Some words may be used more than once.*

| | | |
|---|---|---|
| incident | diffuse | angle of incidence |
| equal | angle of reflection | ray |
| normal | reflected | regular |

1. A single line of light energy is called a _____ray_____.

2. A ray that strikes a surface is called an _____incident_____ ray.

3. A "bounced" ray is called a _____reflected_____ ray.

4. A line that makes a 90° angle to a surface is called a _____normal_____.

5. The angle between an incident ray and its normal is called the _____angle of incidence_____.

6. The angle between a reflected ray and its normal is called the _____angle of reflection_____.

7. An angle of incidence is _____equal_____ to its angle of reflection.

8. There are two kinds of reflections. They are _____regular_____ and _____diffuse_____.

9. A perfectly even surface gives a _____regular_____ reflection.

10. An uneven surface gives a _____diffuse_____ reflection.

**Try this at home**

You can show yourself the difference between regular and diffuse reflection. You will need these materials: a flashlight, a large white sheet of paper, a small, clean mirror, and a small object such as a candle.

Place the mirror up against the sheet of paper. See Figure E. Turn on the flashlight, and turn out the room lights. The room should be dark. Shine the flashlight at the mirror and the sheet of paper, as shown in the figure. The flashlight beam should be at a right angle to the mirror and paper. Make sure that the flashlight is far enough away so that most of the paper is lit. Now, stand off to one side and look at both the mirror and the sheet of paper.

*Look carefully at what you have set up. Try to answer these questions:*

1. Does the paper appear light or dark? _____light_____ Is the paper illuminated?

   _____yes_____ How do you know?

   Because it can be seen, and is reflecting light from the light source.

2. Does the mirror appear light or dark? _____dark_____ Is the mirror illuminated?

   _____yes_____ How do you know?

   Because the candle is reflected in the mirror, showing that it is reflecting light.

3. Is the paper reflecting light? If so, is the reflection regular or diffuse? Yes, diffuse

4. Is the mirror reflecting light? _____yes_____ Is the mirror's reflection regular or

   diffuse? regular

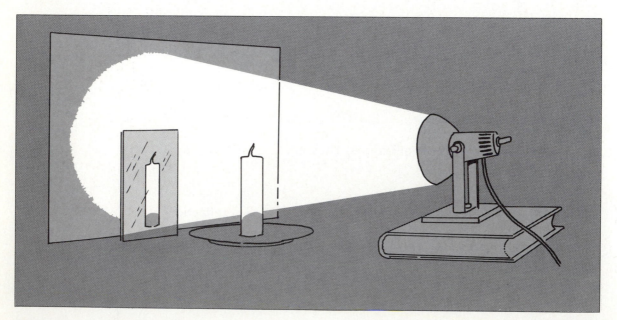

**Figure E**

# What is refraction?

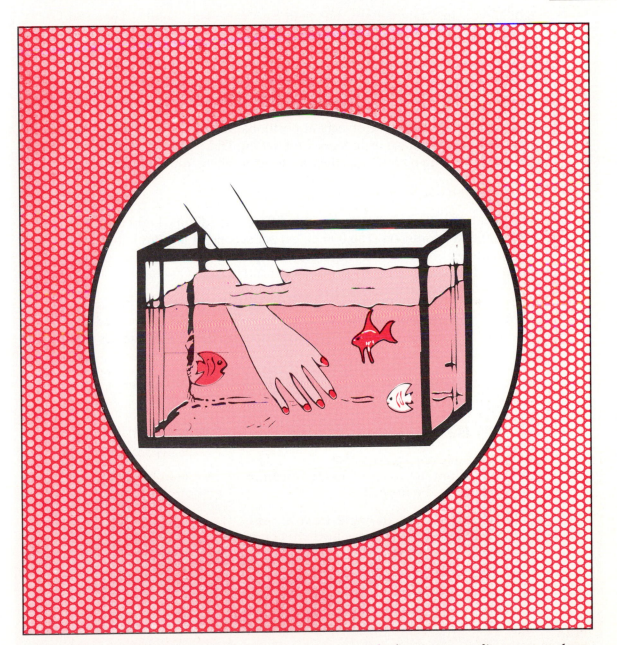

**refraction:**   the bending of light as it passes at an angle from one medium to another
**density:**   the mass of a given volume

# LESSON 12 | What is refraction?

Light travels in straight lines. But light rays can also "bend." They can change direction.

You have seen that light changes direction when it is reflected. Light also changes direction when it passes at an angle from one medium into another medium. This bending is called **refraction** [ree FRAK shun].

Refraction causes us to see objects at positions different from their actual positions. You may have experienced refraction. Did you ever reach into a fish tank to pick up a rock? Was the rock exactly where you thought it was?

How can refraction be explained?

Light travels at different speeds through different mediums. Light travels at about 300,000 kilometers (186,000 miles) per second in air. But light slows down in other substances. In water, for example, light slows down to about 225,000 kilometers (140,000 miles) per second.

The speed at which light travels through a medium depends upon the **density** of that medium. Density has to do with how closely packed the molecules of a substance are. The more closely packed the molecules are, the more dense the substance is.

Different substances have different densities. For example, water is more dense than air.

The following are the Laws of Refraction. They explain how light "bends."

- Light that moves at an angle from a less dense medium to a more dense medium bends towards the normal.

- Light that moves at an angle from a more dense medium to a less dense medium bends away from the normal.

- Light that moves straight on from one medium to another does not bend. It is not refracted.

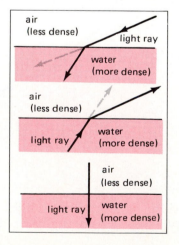

air (less dense)
light ray
water (more dense)

air (less dense)
light ray
water (more dense)

air (less dense)
light ray
water (more dense)

# UNDERSTANDING REFRACTION

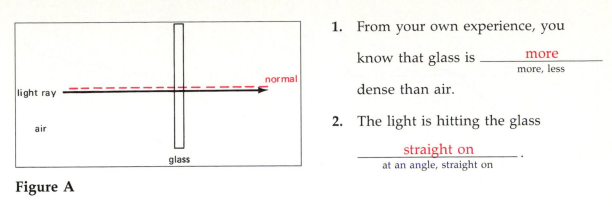

**Figure A**

1. From your own experience, you know that glass is _____more_____
   more, less
   dense than air.

2. The light is hitting the glass
   _____straight on_____.
   at an angle, straight on

3. The light _____is not_____ bending. It _____is not_____ being refracted.
   is, is not        is, is not

4. Why isn't the light being refracted? It is hitting the glass at an angle, not straight on.

5. Write the part of the Law of Refraction that explains why this is happening.

   Light that moves straight on from one medium to another is not refracted.

Look at Figures B through G. In each, light is being refracted. The dotted line in color is the normal. Is the light being refracted towards the normal or away from the normal?

*Complete the sentence under each figure.*

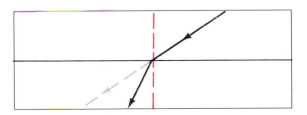

**Figure B**

6. Light is being refracted
   _____towards_____ the normal.
   towards, away from

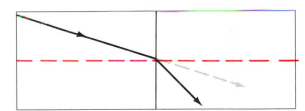

**Figure C**

7. Light is being refracted
   _____away from_____ the normal.
   towards, away from

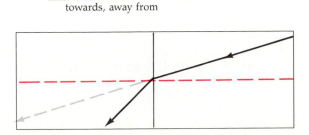

**Figure D**

8. Light is being _____refracted away from_____ the normal.

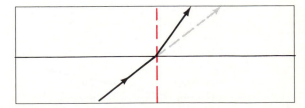

**Figure E**

9. Light _____is being refracted toward the_____ normal.

*Now, answer with complete sentences.*

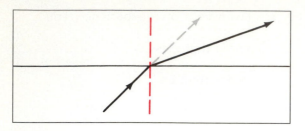

**Figure F**

10.  <u>Light is being refracted away from</u>

<u>the normal.</u>

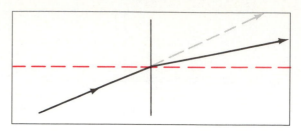

**Figure G**

11.  <u>Light is being refracted toward the</u>

<u>normal.</u>

## MORE ABOUT REFRACTION

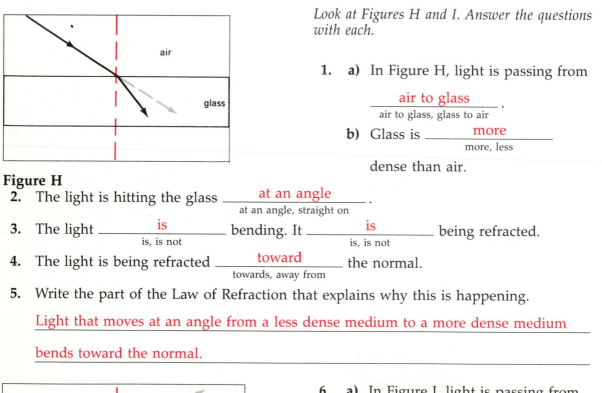

**Figure H**

*Look at Figures H and I. Answer the questions with each.*

1.  **a)** In Figure H, light is passing from

    <u>air to glass</u> .
    air to glass, glass to air

    **b)** Glass is <u>more</u>
    more, less

    dense than air.

2.  The light is hitting the glass <u>at an angle</u> .
    at an angle, straight on

3.  The light <u>is</u> bending. It <u>is</u> being refracted.
    is, is not      is, is not

4.  The light is being refracted <u>toward</u> the normal.
    towards, away from

5.  Write the part of the Law of Refraction that explains why this is happening.

    <u>Light that moves at an angle from a less dense medium to a more dense medium</u>

    <u>bends toward the normal.</u>

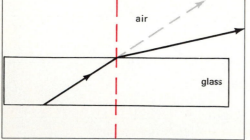

**Figure I**

6.  **a)** In Figure I, light is passing from

    <u>glass to air</u> .
    air to glass, glass to air

    **b)** Air is <u>less</u> dense
    more, less

    than glass.

7.  The light is hitting the air

    <u>at an angle</u> .
    at an angle, straight on

**8.** The light ___is___ bending. It ___is___ being refracted.
is, is not · is, is not

**9.** The light is being refracted ___away from___ the normal.
towards, away from

**10.** Write the part of the Law of Refraction that explains why this is happening.

Light that moves at an angle from a more dense medium to a less dense medium

bends away from the normal.

## REFRACTION AND CHANGE OF POSITION

*Study Figure J. Answer the questions.*

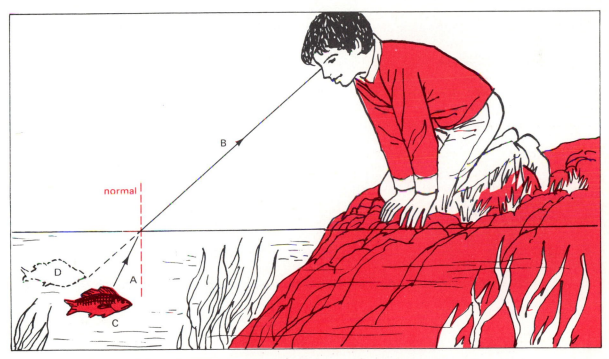

**Figure J** *The fish is actually at C. But to the boy, the fish appears to be at D.*

**1.** The boy sees the fish because light is traveling _____
from the boy's eyes to the fish,

___from the fish to the boy's eyes___ .
from the fish to the boy's eyes

**2.** The fish is ___reflecting light___ .
giving off its own light, reflecting light

**3.** The light is moving from ___water to air___ .
water to air, air to water

**4.** The light from the fish is being refracted ___away from___ the normal.
towards, away from

5. The boy sees the fish in line with the refracted light. The refracted light is

_____B_____ .
   A, B

6. Refraction _____does_____ seem to change the position of an object.
   does, does not

## FILL IN THE BLANK

*Complete each statement using a term or terms from the list below. Write your answers in the spaces provided. Some words may be used more than once.*

| | | |
|---|---|---|
| is not | refraction | more slowly |
| away from | more | air |
| at an angle | toward | less |

1. The bending of light as it passes from one medium to another is called

   _____refraction_____ .

2. Refraction takes place when light strikes a surface _____at an angle_____ to the normal.

3. Light that strikes a surface in the same direction as the normal _____is not_____

   refracted.

4. Light travels at about 300,000 kilometers per second in _____air_____ .

5. Glass and water are _____more_____ dense than air.

6. Light travels _____more slowly_____ in glass or water than it does in air.

7. Light that moves at an angle from a less dense medium to a more dense medium is

   refracted _____toward_____ the normal.

8. Light that moves at an angle from a more dense medium to a less dense medium is refracted _____away from_____ the normal.

9. The light ray in Figure K is being refracted _____away from_____ the normal.

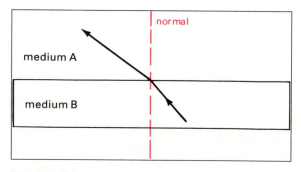

**Figure K**

10. A is _____less_____ dense than B.

Try this at home.
Place a penny into a shallow bowl.

Move back slowly. Stop when you no longer can see the penny.

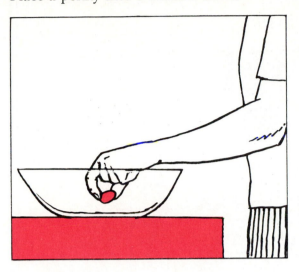

**Figure L**

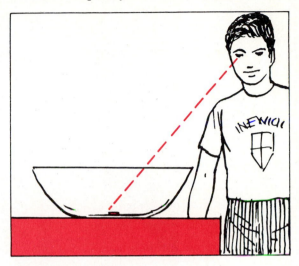

**Figure M**

Have a member of your family slowly pour water into the bowl. (Careful, don't move the penny.)

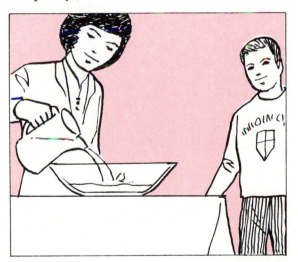

**Figure N**

Notice what happens.

1. Write down what you see. <u>The penny came into view again as water was added to the bowl.</u>

2. Try to explain why this happens. <u>When water was added, light coming from the bowl was refracted away from the normal as it moved from more dense water to less dense air. This caused the image of the penny to appear to the observer.</u>

# SCIENCE *EXTRA*

## Opticians

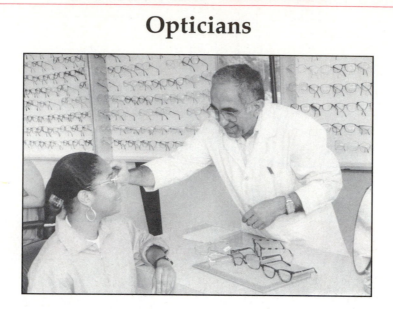

Think of this: *all* people, providing they live beyond the age of 40 or so, will need eyeglasses. The need may be for distance vision only, near vision only, or both.

Now consider this: every pair of eyeglasses required the specialized skills of one or more *opticians.* Opticians are always in demand. If you are mechanically inclined, you may wish to consider a career in this well-paying and growing field.

What do opticians do?

Opticianry is divided into two basic groups—*mechanical optics* and *dispensing.*

- *Mechanical optics,* itself, has two major divisions—*surfacing* and *edging and assembly.*

A *surfacing* optician works on the lens *surface* only. He or she grinds the *curves* of the lens. The curves determine lens power.

An *edging and assembly* optician grinds the *edges* of lenses to a specific size and shape. Then they are inserted into the frame for which they were edged.

- A *dispensing* optician may do mechanical optics. But he or she may also do steps that a mechanical optician may *not* do. A dispensing optician may take facial measurements to determine lens positioning. The dispensing optician may further deliver and adjust eyeglasses to a patient.

There are no specific educational requirements. However, mechanical ability and simple math skills are essential.

Some public as well as private vocational schools offer courses in mechanical optics. Beyond that, at least two years of apprenticeship is needed to gain experience. Often, prior training is not needed for apprenticeship. But one thing certainly makes sense —FINISH HIGH SCHOOL FIRST.

Are you interested in a career as an optician? If you are, see your counselor and get on the right track—*soon.*

# What is the spectrum?

13

**prism:** a special glass that bends light rays.

**spectrum:** a band of radiation that includes different frequencies.

**visible spectrum:** radiation that we can see, including the seven colors of the rainbow

**electromagnetic spectrum:** radiant energy of all frequencies, from radio waves to cosmic rays

# LESSON 13 | What is the spectrum?

Sunlight seems to have no color at all. Yet it is really made up of every color, from red to violet. You can see this for yourself when you see a rainbow. During rain, there are billions of tiny water droplets in the air. When the sun shows, they can break up sunlight into the **spectrum** of colors we call the rainbow.

You can do the same thing with a glass **prism.** This is special glass that breaks up light into the **visible spectrum:** red, orange, yellow, green, blue, indigo, and violet. The order of the colors never changes.

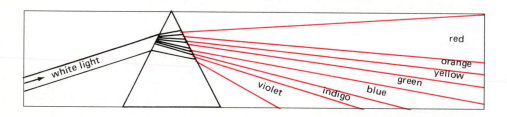

When you studied sound, you learned that pitch depends on the frequency of sound energy. Similarly, the variety of colors depends upon the frequency of the light energy. Each color in the visible spectrum has a different frequency of vibration. Red light has the lowest frequency. As you go from red to violet, frequency increases.

Visible light is just a small part of a broad spectrum of energy. This energy spectrum is called the **electromagnetic** [ih LEK tro mag NET ik] **spectrum,** or E-M spectrum, for short. The other parts of the E-M spectrum are shown in Figure C on page 81. They include radio waves, microwaves, infrared [in fruh RED] waves, ultraviolet rays, x-rays, gamma rays, and cosmic rays. Notice where visible light fits into the E-M family.

# UNDERSTANDING THE VISIBLE SPECTRUM

Only the visible part of the electromagnetic spectrum vibrates at frequencies that the human eye can sense. Each color has its own frequency.

*Study Figure A. Then answer the questions.*

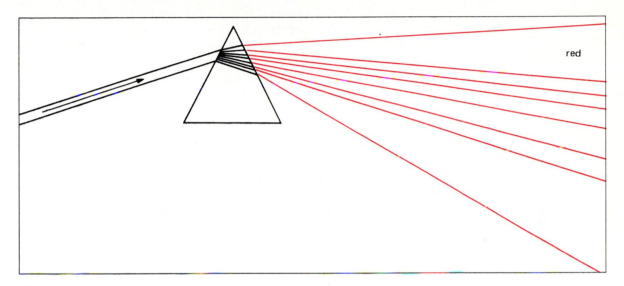

**Figure A**

1. What kind of light is entering the prism? _____ordinary_____

2. The prism is _____refracting_____ the light.
   <u>reflecting, refracting</u>

3. The light is breaking up. It is separating into a rainbow of colors. What do we call

   this rainbow of colors? _____visible spectrum_____

4. Which color is refracted the most? _____violet_____

5. Which color is refracted the least? _____red_____

6. Which color vibrates the fastest? _____violet_____

7. Which color vibrates the slowest? _____red_____

## FILL IN THE BLANK

Complete each statement using a term or terms from the list below. Write your answers in the spaces provided. Some words may be used more than once.

| | | |
|---|---|---|
| visible spectrum | blue | yellow |
| violet | red | how fast |
| many | green | indigo |
| orange | prism | |

1. Ordinary light is really made up of _____ many _____ colors.

2. The colors that make up light are called the _____ visible spectrum _____.

3. The colors of the visible spectrum in order are _____ red _____

    _____ orange _____ _____ yellow _____ _____ green _____ _____ blue _____

    _____ indigo _____ _____ violet _____.

4. Color depends upon _____ how fast _____ light energy vibrates.

5. The color that vibrates the fastest is _____ violet _____.

6. The color that vibrates the slowest is _____ red _____.

7. We can separate the colors of light with a _____ prism _____.

## REACHING OUT

In Figure B, light is passing through two prisms. They are facing in opposite directions.

1. What kind of light do you think is coming out of the prism on the right?

    ordinary light

2. What does this prove? A prism can

    refract the visible spectrum and form

    ordinary light.

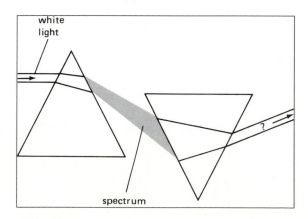

Figure B

# UNDERSTANDING THE ELECTROMAGNETIC SPECTRUM

Each kind of electromagnetic energy vibrates at a different frequency. As you move from left to right in the diagram below, the frequency of vibration increases. The wavelength decreases as you move from left to right.

Our eyes cannot see most parts of the E-M spectrum. But some parts that we cannot see do affect the human body. Infrared energy is the heat we feel from the sun or any hot object. A small amount of ultraviolet energy is healthy for most plants and animals, including humans. But too much ultraviolet can be harmful. Ultraviolet rays can give us sunburn, for example.

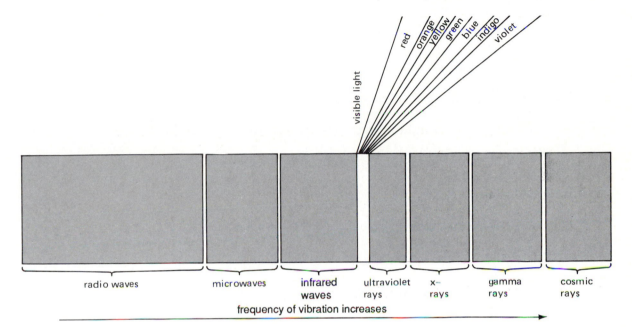

**Figure C**

1. The members of the electromagnetic spectrum vibrate

   ___at different speeds___ .
   
   at one speed, at different speeds

2. The farther to the right you go on the electromagnetic spectrum, the

   ___faster___ the waves vibrate.
   
   faster, slower

3. Which vibrate faster:

   a) gamma rays or radio waves? ___gamma rays___

   b) ultraviolet rays or x-rays? ___x-rays___

   c) infrared waves or cosmic rays? ___cosmic rays___

   d) visible light or microwaves? ___visible light___

4. Violet ___can___ be seen. But ultraviolet ___cannot___ be seen.
   
   can, cannot         can, cannot

5. Red ___can___ be seen. But infrared ___cannot___ be seen.
   
   can, cannot         can, cannot

81

# TRUE OR FALSE

*In the space provided, write "true" if the sentence is true. Write "false" if the sentence is false.*

___True___  1. Visible light is part of the electromagnetic spectrum.

___True___  2. Visible light takes up only a small part of the electromagnetic spectrum.

___False___  3. Every member of the electromagnetic spectrum vibrates at the same speed.

___False___  4. We can see every member of the electromagnetic spectrum.

___False___  5. We can see ultraviolet light.

___False___  6. We can see infrared light.

___True___  7. Ultraviolet light vibrates too quickly for us to see it.

___False___  8. Infrared light vibrates too quickly for us to see it.

___True___  9. Infrared rays are heat rays.

___True___  10. The sun gives off ultraviolet and infrared energy.

# What gives an object its color?

**filter:** a transparent substance that transmits some colors and absorbs others

# LESSON 14 | What gives an object its color?

Joan's dress is red. Tom is wearing a blue shirt. Grass is green. An orange is of course . . . orange. Color, color, everywhere! Look around. How many colors do you see?

What causes color in the objects we see? You have learned that when light strikes an object, it can be reflected, absorbed, or transmitted. The color we see depends upon whether the object is opaque or transparent. The color also depends upon how much of the light is reflected, absorbed, or transmitted.

**OPAQUE OBJECTS** The color of an opaque object is the color that it reflects. For example, a red object reflects only red light. It absorbs all other colors. A green object reflects only green light. It absorbs all other colors.

What about white and black objects? A white object reflects all the colors that make up ordinary light. A black object, on the other hand, absorbs all the colors that strike it. No color is reflected.

Most objects reflect more than one color. The colors combine. We see them as mixtures of colors, like blue-green and red-orange.

**TRANSPARENT OBJECTS** The color of a transparent (or translucent) object is the color that passes through the object. For example, red glass transmits only red light. It absorbs all other colors. Blue glass transmits only blue light. It absorbs all other colors.

Ordinary window glass transmits all colors. Ordinary light that has passed through window glass still has all the colors of ordinary light. None of the colors has been absorbed by the glass.

You have learned that some transparent substances transmit only some colors. All other colors are blocked. These substances are called **filters.** Filters are often used in spotlights and in photography.

# OBJECTS AND THEIR COLORS

*Study Figures A through E. Answer the questions with each.*

## OPAQUE OBJECTS

1. List the colors that are striking the object in Figure A. <u>red, orange, yellow, green,</u> <u>blue, indigo, violet</u>

2. Together, these colors make up what kind of light? <u>ordinary light</u>

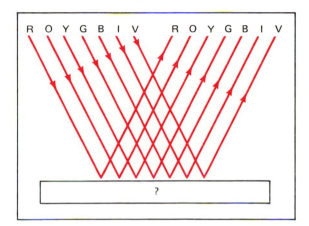

**Figure A**

3. List the colors that are being reflected. <u>red, orange, yellow,</u> <u>green, blue, indigo, violet</u>

4. Together, these colors make up what kind of light? <u>ordinary</u>

5. The color of an opaque object is the color that the object <u>reflects</u> .
   reflects, absorbs, transmits

6. What color is the object in Figure A? <u>white</u>

7. How do you know? <u>The object reflects all colors in the visible spectrum.</u>

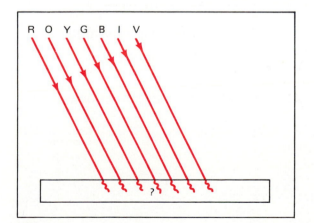

**Figure B**

8. List the colors that are striking the object in Figure B. <u>red, orange,</u> <u>yellow, green, blue, indigo, violet</u>

9. All the colors are being <u>absorbed</u> .
   reflected, absorbed

10. None of the colors is being <u>reflected</u> .
    reflected, absorbed

11. The color of an opaque object is the color that the object <u>reflects</u> .
    reflects, absorbs, transmits

12. What color is the object in Figure B? <u>black</u>

13. How do you know? <u>It reflects no color.</u>

**14.** List the colors that are striking the object in Figure C. <u>red, orange, yellow, green,</u>
<u>blue, indigo, violet.</u>

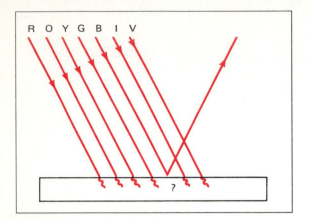

**Figure C**

**15.** List the colors that are being
absorbed. <u>red, orange, yellow,</u>
<u>green, indigo, violet.</u>

**16.** Which color is being reflected?
<u>blue</u>

**17.** What gives an opaque object its
color? <u>The color it reflects</u>

**18.** What is the color of the object in Figure C? <u>Blue</u>

Why? <u>It reflects only blue.</u>

## TRANSPARENT OBJECTS

**19.** List the colors that are striking the glass in Figure D. <u>red, orange, yellow, green,</u>
<u>blue, indigo, violet.</u>

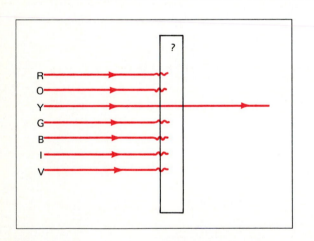

**Figure D**

**20.** Which of these colors are being
absorbed? <u>red, orange, green, blue,</u>
<u>indigo, violet.</u>

**21.** Which color is being transmitted?
<u>yellow</u>

**22.** The color of a transparent object is
the color it <u>transmits</u>.
reflects, absorbs, transmits

**23.** What is the color of the glass in Figure D? <u>yellow</u>

Why? <u>It is transmitting only yellow.</u>

**24.** What is the definition of a filter? _A substance that absorbs some colors and_

_transmits other colors._

**25.** Is this glass a filter? _____yes_____

    Why? _It blocks all colors except yellow._

Light is striking the glass in Figure E.

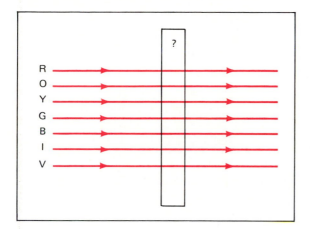

**Figure E**

**26.** Are any colors being absorbed?

    _____No_____

**27.** All the colors are being

    _____transmitted_____ .
    reflected, transmitted

**28.** What kind of light is leaving the

    glass? _____ordinary_____

**29.** Is this glass a filter? _____No_____

    Why? _It transmits all colors in visible light._

**30.** Does this glass have a color? _____no_____

**31.** What kind of glass is this? _____clear_____

**32.** Where can it be found? _____windows_____

## REACHING OUT

The colors of the American flag are red, white, and blue.

What color or colors does each of these colors reflect?

**1.** Red _____red_____

**2.** White _____all the colors_____

**3.** Blue _____blue_____

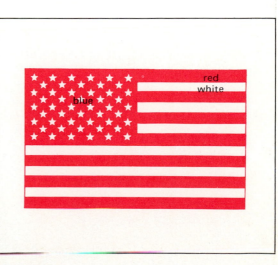

**Figure F**

# FILL IN THE BLANK

*Complete each statement using a term or terms from the list below. Write your answers in the spaces provided. Some words may be used more than once.*

green         all         transparent
filter        red         does not
absorbs       opaque      transmits
reflects

1.  An object that does not allow light to pass through it is said to be

    _____opaque_____ .

2.  An object that does allow light to pass through it is said to be _____transparent_____ .

3.  The color of an opaque object is the color it _____reflects_____ .

4.  A white object reflects _____all_____ the colors that make up white light.

5.  A black object _____does not_____ reflect light. A black object _____absorbs_____
    all the light that strikes it.

6.  A green sweater absorbs all colors except _____green_____ .

7.  The color of a transparent object is the color it _____transmits_____ .

8.  Window glass transmits _____all_____ the colors that strike it.

9.  A red glass transmits only the color _____red_____ . It _____absorbs_____ all
    other colors.

10. A substance that transmits some colors and blocks others is called a

    _____filter_____ .

# What is a lens?

**magnifies:** makes larger
**minifies:** makes smaller
**lens:** a transparent material that refracts light in a definite way
**convex lens:** a lens that is curved outward
**concave lens:** a lens that is curved inward
**image:** visual impression produced by reflection or refraction

# LESSON 15 | What is a lens?

Have you ever snapped a photo or looked through a microscope? If your answer is yes, then you have used an optical device.

There are many kinds of optical devices. Cameras and microscopes are two examples. Others are projectors, binoculars, telescopes, and even eyeglasses.

Every optical device is different. But they all have one thing in common. Each one has at least one **lens.**

What is a lens? A lens is a transparent substance that bends or refracts light in a definite way.

Most lenses are made of glass. Some lenses are made of plastic.

Most lenses have one or two curved surfaces.

There are two main types of lenses: convex [kon VEKS] and concave [kon KAVE].

A **convex lens** is thicker at the center than at the edge. It **magnifies** or makes things look bigger.

A convex lens converges or focuses light rays. The point where the light rays meet is called the focal point.

Light that passes through a convex lens can be focused on a screen or other surface. This forms an **image** of the object that gave the light. Convex lenses are used in projectors and cameras.

A **concave lens** is thinner at the center than at the edge. It **minifies** or makes things look smaller.

A concave lens spreads out light rays. They cannot form an image on a screen.

Concave lenses are often used together with convex lenses. They help the convex lenses give sharper images.

Most eyeglass lenses have combinations of concave and convex curves.

# UNDERSTANDING LENSES

Six lenses are shown in Figure A. Study them. Then answer the questions by writing the correct letters.

a   b   c   d   e   f

**Figure A**

**What You Need To Know:**   Plano means "plane" or "flat."

Which lens or lenses . . .

1.  are thicker at the center than at the edge? _a, c, d_

2.  are thinner at the center than at the edge? _b, e, f_

3.  are concave? _b, e, f_

4.  are convex? _a, c, d_

5.  are plano convex? _a, c_

6.  are plano concave? _b, f_

7.  is double concave? _e_

8.  is double convex? _d_

Which lenses . . .

9.  magnify? _a, c, d_

10.  minify? _b, e, f_

11.  refract light? _all_

12.  converge light? _a, c, d_

13.  spread out light? _b, e, f_

14.  can form an image on a screen? _a, c, d_

15.  cannot form an image on a screen? _b, e, f_

16.  are most important for projectors and cameras? _a, c, d_

Now look at Figure B.

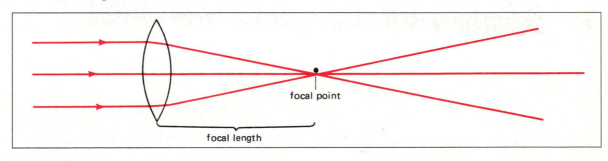

focal point

focal length

**Figure B**

17.  **a)**  Figure B shows a ___convex___ lens.
     <br>concave, convex

**b)**  It ___converges___ light rays.
     <br>converges, spreads

**18.** What do we call the point where light rays converge? _____focal point_____

**19.** What do we call the distance between a lens and its focal point? _____focal length_____

## ABOUT FOCAL LENGTH

Different lenses have different focal lengths.

Focal length depends upon the strength of a lens.

- The stronger the lens, the shorter the focal length.

- The weaker the lens, the longer the focal length.

A strong lens has a deeper curve than a weak lens.

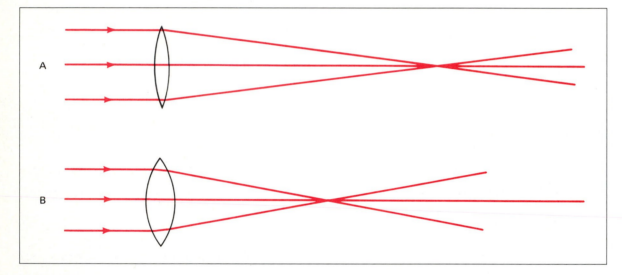

**Figure C**

*Two converging lenses are shown in Figure C. Study the figure. Then answer the questions by writing the correct letters.*

Which lens . . .

1. is more curved? ___B___
2. is less curved? ___A___
3. is stronger? ___B___
4. is weaker? ___A___
5. refracts light less? ___A___

6. refracts light more? ___B___
7. has the shorter focal length? ___B___
8. has the longer focal length? ___A___
9. magnifies more? ___B___
10. magnifies less? ___A___

*Now look at Figure D.*

**Figure D**

**11.** What kind of lens is the boy holding? _____ convex _____

**12.** What kind of lens is the girl holding? _____ concave _____

## FILL IN THE BLANK

*Complete each statement using a term or terms from the list below. Write your answers in the spaces provided. Some words may be used more than once.*

| | | |
|---|---|---|
| refracts | smaller | center |
| focal length | concave | convex |
| edge | focal point | larger |

**1.** A lens is a transparent material that _____ refracts _____ light in a definite way.

**2.** The two main types of lenses are _____ concave _____ and _____ convex _____ lenses.

**3.** A concave lens makes things look _____ smaller _____ .

**4.** A convex lens makes things look _____ larger _____ .

**5.** The thickest part of a convex lens is its _____ center _____ .

**6.** The thickest part of a concave lens is its _____ edge _____ .

**7.** A _____ convex _____ lens can form an image on a screen.

**8.** A _____ concave _____ lens cannot form an image on a screen.

**9.** The point where converging light meets is called the _____ focal point _____ .

**10.** The distance between a lens and its focal point is called its _____ focal length _____ .

## MATCHING

Match each term in Column A with its description in Column B. Write the correct letter in the space provided.

### Column A

<u>c</u>    **1.** focal point

<u>a</u>    **2.** convex

<u>b</u>    **3.** prism

<u>d</u>    **4.** concave

### Column B

**a.** lens that is thicker at the center

**b.** glass shaped like a wedge

**c.** where light rays converge

**d.** lens that is thinner at the center

## REACHING OUT

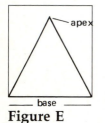

**Figure E**

A prism is shaped like a wedge.

Its point is called the apex. The opposite flat surface is called the base.

Lenses can be described as combinations of prisms.

One kind of lens can be described as a combination of prisms touching at their apexes.

Another kind of lens can be described as a combination of prisms touching at their bases.

**1.** How would you describe a convex lens? <u>A combination of prisms touching at their bases.</u>

**2.** How would you describe a concave lens? <u>A combination of prisms touching at their apexes.</u>

**Figure F**

# How do we see?

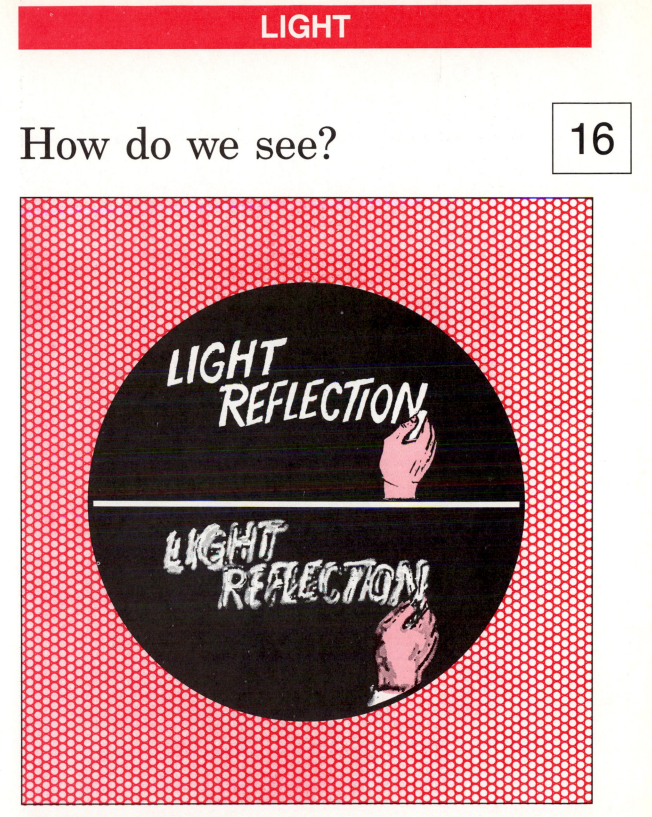

**retina:**  the nerve layer of the eye
**rods:**  nerve cells that are sensitive to brightness
**cones:**  nerve cells that are sensitive to color
**optic nerve:**  nerve that connects the eye to the brain
**lens:**  a refracting part of the eye that changes shape to focus light rays

# LESSON 16 | How do we see?

At this moment you are reading. Your eyes are open. They must be open. Otherwise, you could not see this page—or anything else.

The eyes are sense organs. They are the organs that allow us to see. The eyes receive and focus light. The light causes nerve signals that go to the brain. The brain tells us what the light means. It tells us what we are seeing.

The eye has several transparent parts. Each part refracts light that enters the eye. In a normal eye, the light rays converge exactly upon the **retina** [RET nuh].

The retina is at the back part of the eye. It is made up of two kinds of nerve cells: **rods** and **cones.**

- Rods are sensitive to brightness but not to color.

- Cones are sensitive to color. Without cones, we would not see color. Everything would be seen as black and white and shades of gray.

The tissues of the retina join to form the **optic nerve.** The optic nerve leads into the brain.

Every part of the eye that refracts light has a "set" focus. Its power does not change. Every part, that is, except the **lens.**

The lens of the eye can change focus. It becomes stronger when we are looking at something close-up. This is important because the eye needs a stronger power for close vision. If the lens of the eye did not change focus, close-up things would seem blurry to most people.

# THE EYE

Figure A shows the inside of an eye. It also shows light rays passing through.
*Study the figure. Then answer the questions.*

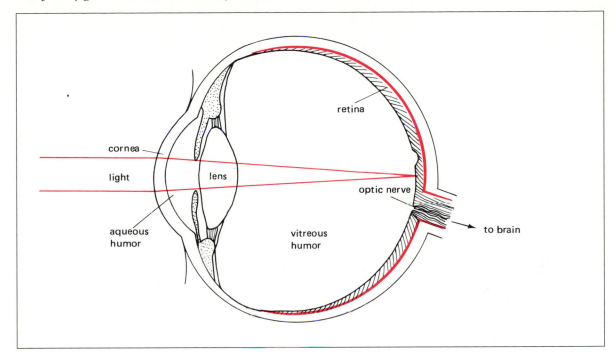

cornea

light

lens

aqueous
humor

retina

optic nerve

vitreous
humor

to brain

**Figure A**

1. The first refracting medium of the eye is sometimes called the "window of the eye."

   What is its name? _____cornea_____

2. Directly behind the cornea is a transparent liquid. What is its name?

   _____aqueous humor_____

3. The eye has a thick double-convex tissue. What is it called? _____lens_____

4. Behind the lens is a transparent jelly-like material. What is its name?

   _____vitreous humor_____

5. List (in order) the refracting parts of the eye:

   _____cornea_____  _____aqueous humor_____  _____lens_____  _____vitreous humor_____

6. In a normal eye, light rays converge upon the _____retina_____ .

7.  The retina is made up of two kinds of nerve cells. Name them.

    _____rods_____ and _____cones_____

8.  Rods are sensitive to _____brightness_____ .

9.  Cones are sensitive to _____color_____ .

10. Retina tissues join to form the _____optic nerve_____ .

11. The optic nerve leads into the _____brain_____ .


## FILL IN THE BLANK

*Complete each statement using a term or terms from the list below. Write your answers in the spaces provided. Some words may be used more than once.*

| | | |
|---|---|---|
| color | converge | refract |
| lens | eye | brightness |
| optic nerve | brain | transparent |
| retina | stronger | |

1.  The organ that is sensitive to light is the _____eye_____ .

2.  The eye has several _____transparent_____ parts.

3.  The transparent parts of the eye bend, or _____refract_____ , light.

4.  The nerve layer of the eye is called the _____retina_____ .

5.  In a normal eye, light rays _____converge_____ upon the retina.

6.  The retina is made up of rods and cones. Rods are sensitive to _____brightness_____ .

    Cones are sensitive to _____color_____ .

7.  Retina tissues join to form the _____optic nerve_____ .

8.  The optic nerve leads into the _____brain_____ .

9.  The part of the eye that can change focus is the _____lens_____ .

10. When we look at close-up things, the power of the lens becomes

    _____stronger_____ .

# MATCHING

Match each term in Column A with its description in Column B. Write the correct letter in the space provided.

### Column A

_____d_____ **1.** eye

_____a_____ **2.** cornea

_____c_____ **3.** retina

_____e_____ **4.** lens of the eye

_____b_____ **5.** optic nerve

### Column B

**a)** first refracting part of the eye

**b)** leads to the brain

**c)** made up of rods and cones

**d)** organ of sight

**e)** can change its power

## IDENTIFYING PARTS OF THE EYE

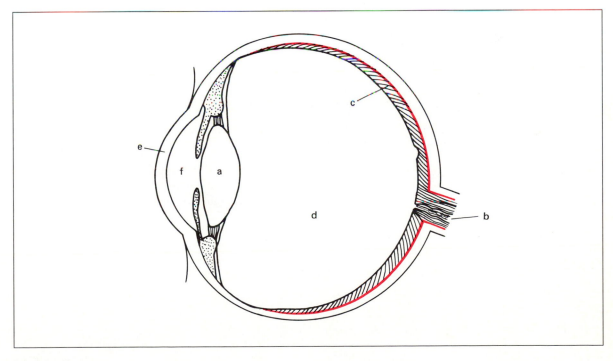

**Figure B**

Study Figure B. Then, identify these parts of the eye by letter.

_____b_____ **1.** optic nerve      _____f_____ **4.** aqueous humor

_____e_____ **2.** cornea      _____a_____ **5.** lens

_____c_____ **3.** retina      _____d_____ **6.** vitreous humor

99

Five more parts of the eye are listed below. Can you locate them by their descriptions?

*Write the correct names next to the numbers on Figure C.*

**Choroid**    Middle layer of the eye. Rich in blood vessels. Supplies the eye with food and oxygen.

**Ciliary Muscle**    Tiny muscle that changes the shape and power of the lens of the eye.

**Pupil**    Opening of the eye through which light enters.

**Iris**    Gives an eye its color. Opens wider or narrower depending upon the amount of light present.

**Sclera**    Tough, white outer layer of the eye.

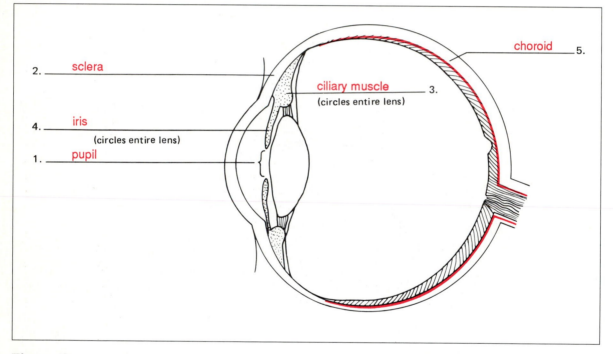

**Figure C**

# How do eyeglasses help some people to see better?

17

**farsightedness:** blurred vision caused when light rays converge beyond the retina
**nearsightedness:** blurred vision caused when light rays converge in front of the retina

# LESSON 17 | How do eyeglasses help some people to see better?

How many people do you know who wear eyeglasses? Probably many. Eyeglasses are very common. In the United States alone, more than 100 million people wear them.

<u>Why</u> are eyeglasses worn?

Some people do not see clearly. They have blurred vision. In most cases, eyeglasses can clear the blur.

<u>What</u> causes blurred vision?

<u>How</u> do eyeglasses help?

Let us first see what is meant by a "normal" eye.

A normal eye sees clearly. Its length—that is, the distance from the cornea to the retina—is just right. Light rays that enter a normal eye converge directly upon the retina. The image, therefore, is in perfect focus. No correcting lens is needed.

In some eyes, however, the length of the eyeball is not right. It is either a bit too long or a bit too short. Because of this, light rays do not converge on the retina. The image is out of focus. Vision is blurred.

There are two main types of blurred vision: **nearsightedness** and **farsightedness.** Both may be corrected with lenses. The lenses bend or refract the light rays so that they converge on the retina.

NEARSIGHTEDNESS

*CAUSE* • A nearsighted eye is slightly <u>longer</u> than normal. Light rays converge at a point in <u>front</u> of the retina.

*CORRECTION* • Nearsightedness may be corrected with a <u>concave</u> lens.

FARSIGHTEDNESS

*CAUSE* • A farsighted eye is slightly <u>shorter</u> than normal. Light rays converge at a point <u>beyond</u> the retina.

*CORRECTION* • Farsightedness may be corrected with a <u>convex</u> lens.

## TYPES OF LENSES

Study Figure A. Then answer the questions.

1. Which lens is <u>concave</u>? ___b___
   a, b

2. Which lens is <u>convex</u>? ___a___
   a, b

3. A convex lens makes light rays

   ___come together___.
   come together, spread apart

4. A concave lens makes light rays

   ___spread apart___.
   come together, spread apart

Remember these facts. You will need to know them for the next exercises.

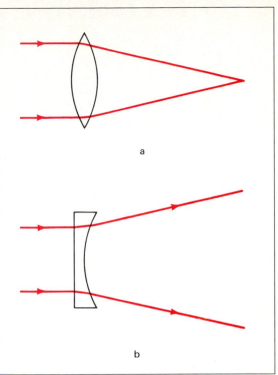

**Figure A**

## TYPES OF VISION

Figures B through D show three types of vision. Study the figures. Then answer the questions with each.

1. "Eyeball length" means the distance between the ___cornea___ and

   ___retina___.

Look at Figure B.

2. a) Is a normal eye too long? ___no___

   b) Is a normal eye too short? ___no___

3. In a normal eye, light rays converge

   ___directly on___ the
   directly on, in front of, beyond

   retina.

4. The image on the retina is

   ___in focus___.
   in focus, out of focus

5. Vision is ___clear___.
   blurred, clear

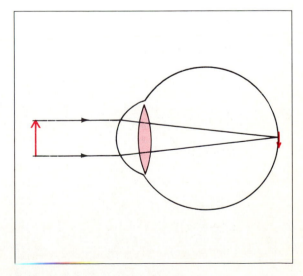

**Figure B**  *The normal eye*

103

6. Compare the eye in Figure C to the normal eye. A nearsighted eye is

   _____longer_____ than a normal eye.
   <sub>longer, shorter</sub>

7. Light rays converge _____
   <sub>directly upon,</sub>

   ____in front of____ the retina.
   <sub>in front of, beyond</sub>

8. The image on the retina is

   _____out of_____ focus.
   <sub>in, out of</sub>

9. Vision is _____blurred_____ .
   <sub>blurred, clear</sub>

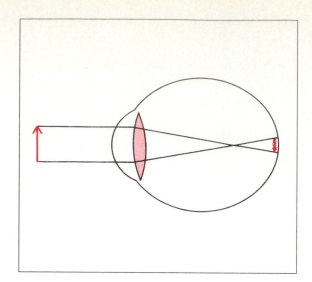

**Figure C**  *The nearsighted eye*

10. **a)** To clear the blur, we must make the light rays fall directly upon the retina. The

    rays must be made to _____spread apart_____ before they reach the eye.
    <sub>(spread apart, converge)</sub>

    **b)** What kind of lens does this? _____concave_____
    <sub>concave, convex</sub>

11. What kind of lens can correct nearsightedness? _____concave_____
    <sub>concave, convex</sub>

12. Compare the eye in Figure D to the normal eye. A farsighted eye is

    _____shorter_____ than a normal
    <sub>longer, shorter</sub>

    eye.

13. Light rays converge

    _____beyond_____ the
    <sub>directly on, in front of, beyond</sub>

    retina.

14. The image on the retina is

    _____out of_____ focus.
    <sub>in, out of</sub>

15. Vision is _____blurred_____ .
    <sub>blurred, clear</sub>

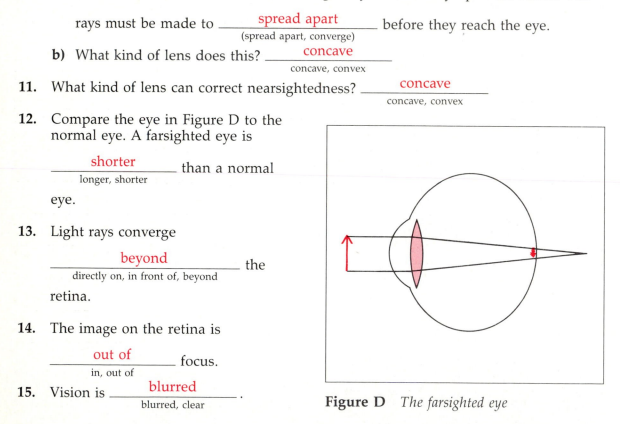

**Figure D**  *The farsighted eye*

16. To clear the blur, we must make the light rays fall directly upon the retina. The rays must be made to _____converge_____ before they reach the eye.
    <u>(spread apart, converge)</u>

17. What kind of lens does this?

    _____convex_____
    concave, convex

18. What kind of lens can correct farsightedness? _____convex_____

## COMPLETE THE CHART

*Several characteristics are listed on the chart. Each one is a characteristic of a normal eye, or a nearsighted eye, or a farsighted eye. Decide which column(s) each characteristic fits. Check the correct box(es). (Three of these may be checked in two boxes.)*

|    |                                      | Normal Eye | Nearsighted Eye | Farsighted Eye |
|----|--------------------------------------|:----------:|:---------------:|:--------------:|
| 1. | eyeball too long                     |            | ✔               |                |
| 2. | vision blurred                       |            | ✔               | ✔              |
| 3. | may be corrected with convex lenses  |            |                 | ✔              |
| 4. | light rays converge beyond the retina|            |                 | ✔              |
| 5. | may be corrected with concave lenses |            | ✔               |                |
| 6. | lens correction needed               |            | ✔               | ✔              |
| 7. | eyeball too short                    |            |                 | ✔              |
| 8. | vision perfectly clear               | ✔          |                 |                |
| 9. | image on retina out of focus         |            | ✔               | ✔              |
| 10.| light rays converge in front of retina|           | ✔               |                |
| 11.| image on retina in focus             | ✔          |                 |                |
| 12.| light rays converge upon the retina  | ✔          |                 |                |
| 13.| no lens correction needed            | ✔          |                 |                |

**Figure E**

The eye converges light. However, stronger converging power is needed for close vision than for far vision.

The ciliary [SILL ee er ee] muscle controls the shape and power of the lens. The lens becomes stronger when we are looking at close-up things.

The ciliary muscle stays flexible and does its job well—up to the age of about 40 years. After this age, it becomes sluggish. Because of this, the lens does not change its power as much as is needed. Outside help in the form of eyeglasses may be needed.

The person in Figure E is over 40 years old. She does not need eyeglasses for far vision. But she does need "reading glasses."

What kind of lens would be used in her reading glasses, convex or concave?

_____convex_____

## WORD SEARCH

*The list on the left contains words that you have used in this Lesson. Find and circle each word where it appears in the box. The spellings may go in any direction: up, down, left, right, or diagonally.*

RODS
CONES
CONVERGE
FOCUS
RETINA
CORNEA
LENS
OPTIC
VISION

```
E G R E V N O C F
R Y E C O N O O S
O P T I C R C N E
D N I E N U E E S
S R N E S L A S I
C O A V I S I O N
```

# What is laser light?

18

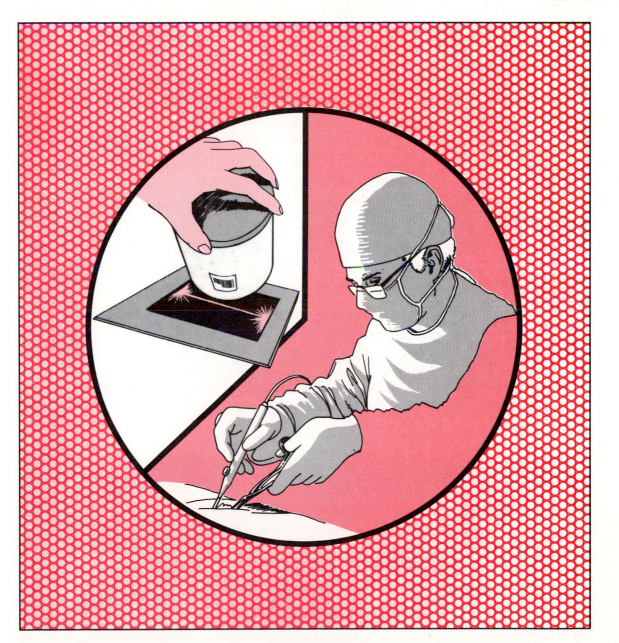

**laser:**  very strong, concentrated, single-color light

# LESSON 18 | What is laser light?

Light is a form of energy. Light can make things move. It can do work for us.

In Lesson 13 you learned that a mixture of all the colors produces white. Sunlight is a mixture of visible light waves that have many wavelengths. These are the waves that our eyes see as colors.

Waves have energy. However, the energy of a mixture of different waves cancels each other out. Single-color light has just one wavelength. It can have more energy than mixed light.

We can produce single-color light by using filters. But filters reduce the amount of light. Very little light energy is left. And what is left spreads out—just like ordinary light does.

Scientists use electronics to produce single-color light. But it is more than just "single color." This light has very special properties.

• This light is very concentrated.

• All the waves are in <u>phase</u>. This means that the crests of all the waves match up with one another. The waves work together.

• The waves stay together. They hardly spread out at all.

This produces an enormous amount of light energy. We call this light **laser** light.

Laser light has many uses, but it is very difficult to handle. Much research is being done to put the laser to work. There has been some success. Laser energy is used in medicine, industry, and scientific research. It is also used by the military for national defense.

Some present-day and future uses of laser energy are shown on the following pages.

# COMPARING ORDINARY LIGHT AND LASER LIGHT

*Figure A shows light from an electric bulb. Figure B shows a beam of laser light. Look at the figures. Then answer the questions.*

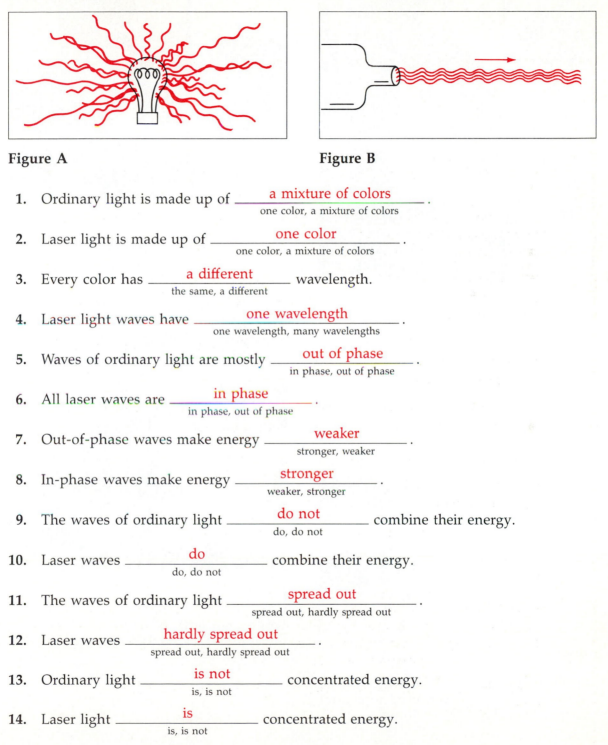

**Figure A**                                   **Figure B**

1. Ordinary light is made up of ___a mixture of colors___ .
   <br>one color, a mixture of colors

2. Laser light is made up of ___one color___ .
   <br>one color, a mixture of colors

3. Every color has ___a different___ wavelength.
   <br>the same, a different

4. Laser light waves have ___one wavelength___ .
   <br>one wavelength, many wavelengths

5. Waves of ordinary light are mostly ___out of phase___ .
   <br>in phase, out of phase

6. All laser waves are ___in phase___ .
   <br>in phase, out of phase

7. Out-of-phase waves make energy ___weaker___ .
   <br>stronger, weaker

8. In-phase waves make energy ___stronger___ .
   <br>weaker, stronger

9. The waves of ordinary light ___do not___ combine their energy.
   <br>do, do not

10. Laser waves ___do___ combine their energy.
    <br>do, do not

11. The waves of ordinary light ___spread out___ .
    <br>spread out, hardly spread out

12. Laser waves ___hardly spread out___ .
    <br>spread out, hardly spread out

13. Ordinary light ___is not___ concentrated energy.
    <br>is, is not

14. Laser light ___is___ concentrated energy.
    <br>is, is not

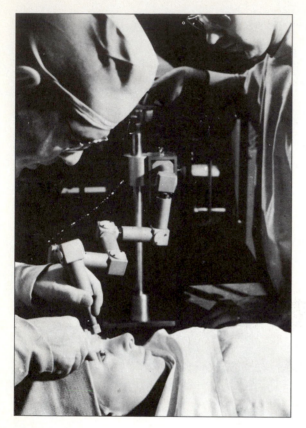

**Figure C**

Lasers are used in delicate eye surgery. They can repair detached (split) retinas.

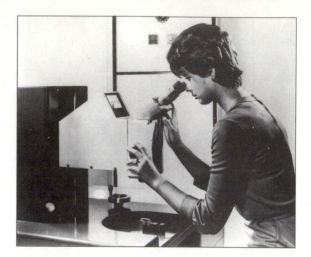

**Figure D**

Lasers are used for drilling and welding ultra-thin holes. A microscope is needed to do these delicate jobs.

**Figure E**

Lasers control machine tools with extreme accuracy. Lasers are even used to control the cutting of cloth as in Figure E.

**Figure F**

Lasers can carry telephone messages. In theory, one laser system can carry 80 million conversations.

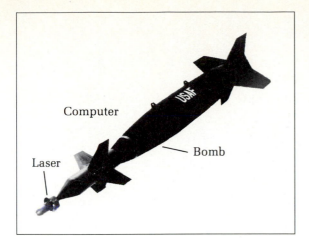

**Figure G**

Laser beams very accurately guide bombs and missiles to targets.

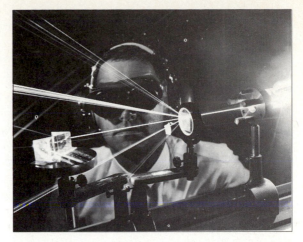

**Figure H**

Scientists use lasers to "look" inside molecules.

Lasers are used in cancer research. They are even used to predict earthquakes.

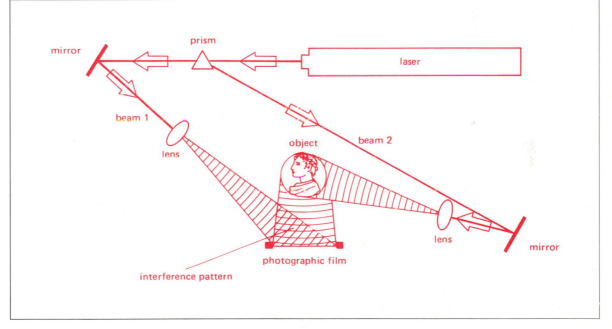

**Figure I**

One of the most exciting uses of laser energy is in holography [huh LOG ruh fee]. In holography, laser beams are used to project a picture of an object. The picture (image) is called a hologram. A hologram looks three-dimensional. A hologram can make an object look as if it is suspended in mid-air. The image shows full roundness and depth. And you can see what is lying behind the object simply by moving your head.

Holography may give us 3-D color movies and TV someday.

## TRUE OR FALSE

*In the space provided, write "true" if the sentence is true. Write "false" if the sentence is false.*

__False__     **1.** Ordinary light is made up of one color.

__True__     **2.** Ordinary light has many wavelengths.

__False__     **3.** The waves of ordinary light are mostly in phase.

__True__     **4.** Laser light is made up of one color.

__True__     **5.** Laser light has only one wavelength.

__True__     **6.** Laser waves are all in phase.

__True__     **7.** Ordinary light spreads out.

__False__     **8.** Laser light spreads out a great deal.

__False__     **9.** Ordinary light produces very strong energy.

__False__     **10.** Laser light produces very weak energy.


## REACHING OUT

The word LASER is made of the first letters of five words. These words describe how a laser is produced. What are these words? (Hint: You may want to do some research.)

__LIGHT__      __AMPLIFICATION__  by __STIMULATED__ __EMISSION__     of __RADIATION__

# What is static electricity? | 19

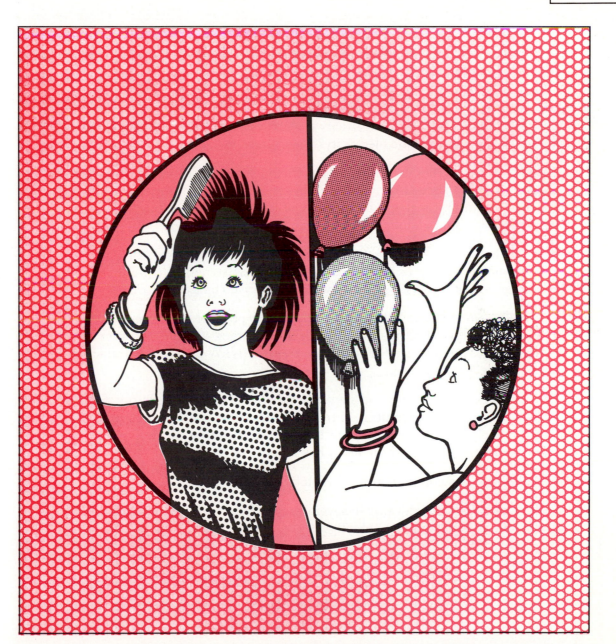

**atom:** the smallest part of an element that has all of the characteristics of that element
**neutral:** having no electrical charge
**friction:** the rubbing of one thing against another thing
**static:** not moving

# LESSON 19 | What is static electricity?

Did you ever walk across a rug, touch something, and get a shock? That shock was caused by **static electricity** [STAT ik i leck TRISS it ee]. Static means not moving. Static electricity is electricity that is not moving along a path. What causes static electricity?

To understand what causes static electricity, you have to know about the **atom.** Scientists have learned that all matter is made up of tiny parts called atoms. An atom is the smallest part of an element that has all of the properties of that element.

Atoms have charges of electrical energy. There are two kinds of charges. There are <u>positive</u> (plus or +) charges. There are also <u>negative</u> (minus or −) charges. An atom has both positive and negative charges.

Usually, an atom has the same number of positive charges as it has negative charges. The positive and negative charges cancel each other out. The charges are balanced. The atom is **neutral** [NEW trul]. A neutral atom has no electrical charge.

Sometimes, the positive and negative charges of an atom are not equal. Then the atom is not neutral. If the atom has more positive charges than negative charges, the whole atom has a positive charge. If there are more negative charges, the whole atom has a negative charge.

<u>Matter that has charged atoms has static electricity.</u>

Static electricity can develop in several ways. One way is by rubbing certain substances together. The rubbing of one object against another object is called **friction** [FRIK shun]. Static electricity is sometimes called friction electricity.

Static electricity is not the same as the electricity we use for light bulbs, motors, toasters and other electrical appliances.

# PLUS AND MINUS CHARGES

Charged matter may have a plus (+) charge or a minus (−) charge.

- Opposite charges attract.

- A plus or minus charge
        and
  a neutral charge also attract.

- Same charges repel.

Four of these pairs will attract. Two pairs will repel.

Which pairs will attract?
Which pairs will repel?
Write your answers below.

+ and +
+ and −
− and −
− and +
neutral and +
neutral and −

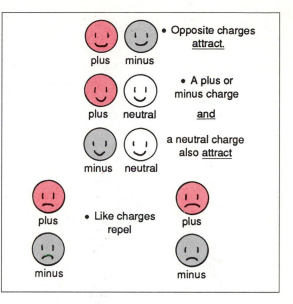

**Figure A**

| ATTRACT | REPEL |
|---|---|
| + and − | + and + |
| − and + | − and − |
| neutral and + | |
| neutral and − | |

A balloon rubbed with a flannel cloth will stick to the cloth.

Do the balloon and the cloth have static electricity? ___yes___

If so, do they have like charges, or opposite charges?
They have opposite charges

A second balloon is rubbed with the same flannel cloth. This balloon also sticks to the cloth.

Do the first balloon and the second balloon have like charges, or opposite charges?
They have like charges

If the charge on the flannel cloth is positive, what is the charge on the two balloons?
negative

*First do step 1. Then do step 2. Answer the questions next to each step.*

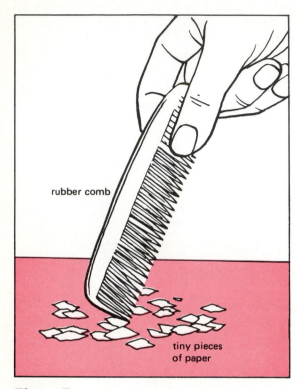

**Figure B**

## STEP 1

Touch a rubber comb to a few tiny pieces of paper. (See Figure B.)

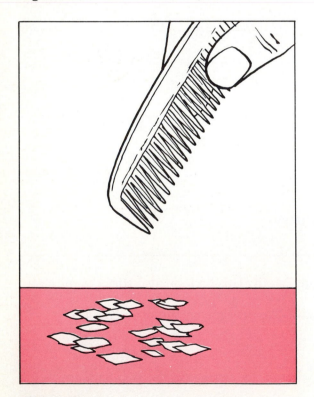

**Figure C**

Then lift the comb. (See Figure C.)

1.  The comb _____**does not**_____

    <sub>does, does not</sub>

    pick up the paper.

2.  The comb _____**is not**_____

    <sub>is, is not</sub>

    charged.

3.  The paper _____**is not**_____

    <sub>is, is not</sub>

    charged.

4.  This shows that objects with no

    charge _____**do not**_____

    <sub>do, do not</sub>

    attract each other.

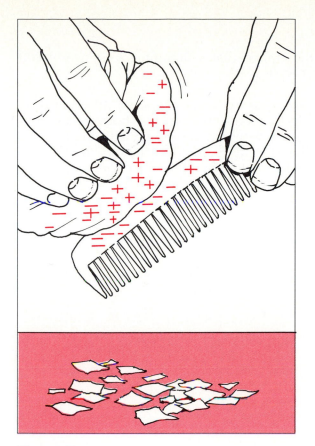

**Figure D**

Rub the comb with a piece of cloth or fur. [Combing your hair may also do the job.] This rubbing causes negative charges to move from the cloth to the comb. (Figure D.)

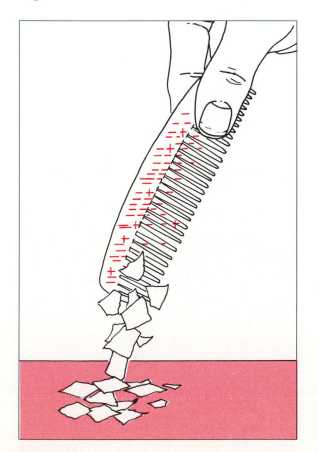

**Figure E**

Touch the comb to the pieces of paper. Then lift the comb. (Figure E.)

1. The comb _____does_____
   does, does not

   pick up the paper.

2. The comb _____has_____
   has, has not

   become charged.

3. The comb now _____has a minus charge_____
   a) has a plus charge.
   b) has a minus charge.
   c) is neutral.

4. The paper _____is neutral_____
   a) has a plus charge.
   b) has a minus charge.
   c) is neutral.

5. This shows that a charged object

   _____does_____ attract a
   does, does not

   neutral object.

117

# UNDERSTANDING LIGHTNING

Lightning is dangerous and spectacular. In the United States, lightning kills nearly 400 people every year, and injures many more. But what is it?

Lightning is a spark that jumps from cloud to cloud or from cloud to earth. It's caused by static electricity that builds up in clouds. Static electricity travels along the shortest path from one point to another. So lightning moving from a cloud toward the ground will tend to strike a tall object. Many houses have lightning rods that stick up higher than the roof. If lightning strikes the rod, the electricity will travel through the metal rod all the way into the ground. No one gets hurt, and the house is not damaged.

### Lightning Safety Rules

During a lightning storm . . .

1.  DON'T run onto an open field.

2.  DON'T stay under a tree.

3.  DO stay indoors or find a place indoors.

4.  If you are in a car during a lightning storm, DO stay there. [Can you figure out why?]

5.  If you are swimming, DO get out of the water.

## WHAT DOES THE PICTURE SHOW?

*Look at the picture. Then answer the questions.*

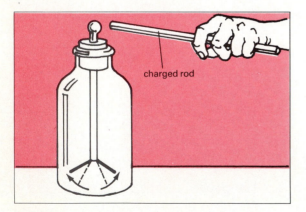
charged rod

**Figure F**

An electroscope is a simple instrument. It tells us if an object has static electricity.

If you hold a charged object near the tip of an electroscope, the leaves move apart.

The leaves move apart because they have

_____the same_____ charges.
the same, opposite

118

## MULTIPLE CHOICE

*In the space provided, write the letter of the phrase that best completes each statement.*

_____c_____    **1.** An atom has

     **a)** only plus charges.   **b)** only minus charges.

     **c)** plus and minus charges.   **d)** no charges.

_____a_____    **2.** Usually, an atom has

     **a)** the same number of plus and minus charges.

     **b)** more plus charges than minus charges.

     **c)** more minus charges than plus charges.   **d)** no charges.

_____b_____    **3.** "Neutral" charge means

     **a)** plus charge.   **b)** no charge.

     **c)** minus charge.   **d)** two plus charges.

_____c_____    **4.** Charged matter has

     **a)** no electricity.   **b)** moving electricity.

     **c)** static electricity.   **d)** only plus charges.

_____b_____    **5.** Static electricity

     **a)** moves in a path.   **b)** does not move in a path.

     **c)** is neutral.   **d)** has only minus charges.

_____c_____    **6.** To make 100 minus charges neutral, you need

     **a)** 50 minus charges and 50 plus charges.   **b)** 100 minus charges.

     **c)** 100 plus charges.   **d)** 50 plus charges.

_____b_____    **7.** Same charges

     **a)** attract.   **b)** repel.   **c)** do not attract or repel.   **d)** attract and repel.

_____a_____    **8.** Opposite charges

     **a)** attract.   **b)** repel.   **c)** do not attract or repel.   **d)** attract and repel.

_____b_____    **9.** Static electricity can come from

     **a)** batteries.   **b)** friction.   **c)** not moving.   **d)** wire.

# MATCHING

*Match each term in Column A with its description in Column B. Write the correct letter in the space provided.*

| | Column A | Column B |
|---|---|---|
| __c__ | 1. opposite charges | a) means "not moving" |
| __d__ | 2. neutral | b) repel |
| __e__ | 3. rubbing | c) attract |
| __a__ | 4. static | d) charges are balanced |
| __b__ | 5. same charges | e) can cause static electricity |

# REACHING OUT

Benjamin Franklin was a famous American. He discovered that lightning is a spark of electricity. Once, during a thunderstorm, he flew a kite with a wire attached to it. He observed that the metal wire attracted lightning.

1. Why should you not do this? __Because lightning contains a dangerous amount of electricity, enough to kill a person.__

2. What can a kite act as? __The kite may act as a conductor of electricity.__

**Figure G**

**Figure H**

# What is electric current? 20

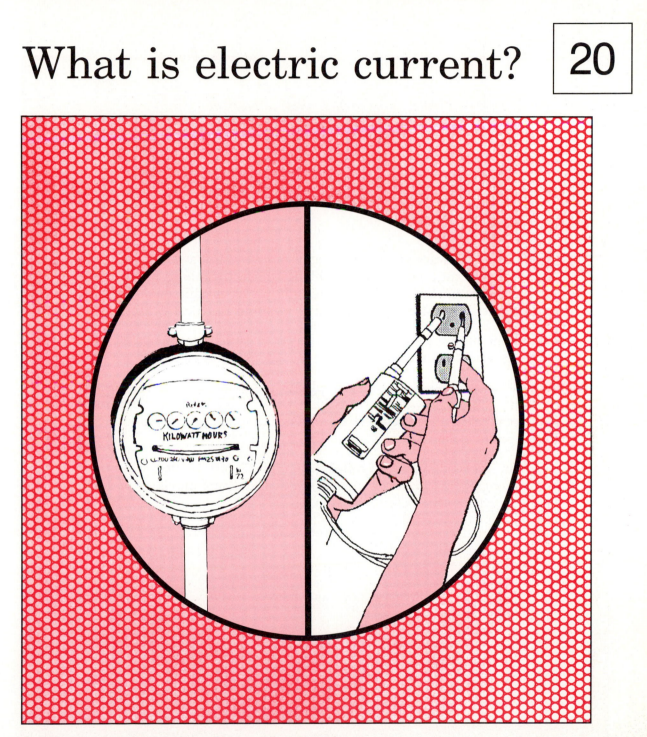

**circuit:** a path that ends at the same point where it starts
**generator:** machine that makes electricity
**electrons:** negatively charged particles in the atom.

# LESSON 20 | What is electric current?

Think of all the ways you use electricity each day. You awake to an alarm clock or radio, turn on an electric light, use an electric toothbrush, or make toast. You watch television, listen to records, use air conditioners. Just think about lights. Almost every place you go you find electrical lighting.

About one hundred years ago, there was no electricity in homes, schools, factories, and offices. Try to imagine your life without electricity!

The electricity that works all your electrical appliances is called **electric current.** This is a flow of **electrons** [i LECK tronz]. Electrons are the parts of the atom that have a negative charge. There is another part of the atom that has a positive charge.

Electrons move along a path called a **circuit** [SIR cut]. While the electrons are moving, the circuit is complete. If the electrons stop moving, the circuit is incomplete and the electricity stops.

Some of our electricity comes from batteries. Small batteries like those used for flashlights are called dry cells. Most of our electricity comes from machines called **generators** [JEN uh ray terz].

Each year, the world uses more and more electricity. More and more generators are needed.

# SOME COMMON ELECTRICAL SYMBOLS

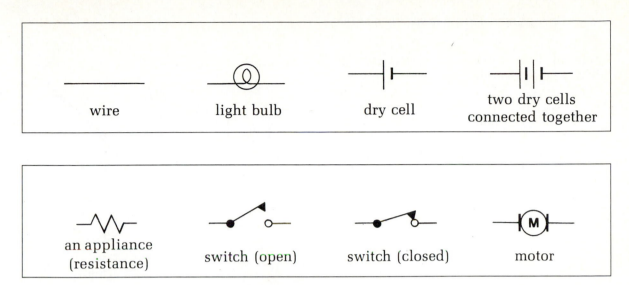

| wire | light bulb | dry cell | two dry cells connected together |

| an appliance (resistance) | switch (open) | switch (closed) | motor |

**Figure A**

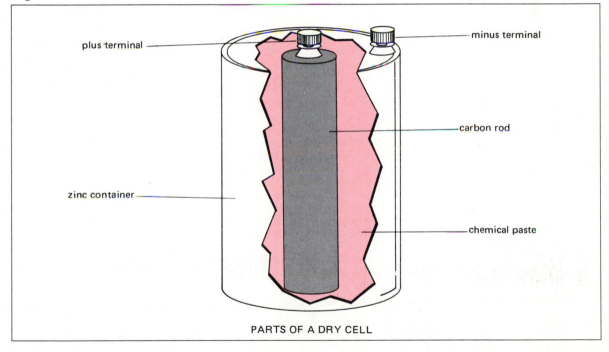

plus terminal — minus terminal

carbon rod

zinc container

chemical paste

PARTS OF A DRY CELL

**Figure B**

## PARTS OF A DRY CELL

A dry cell changes <u>chemical energy</u> to <u>electric energy</u>.

Dry cells come in many different sizes and strengths.

*Look at each picture. Then answer the questions.*

Anything that works with electricity is called an electrical device.

We call some electrical devices appliances. Electricians call them loads.

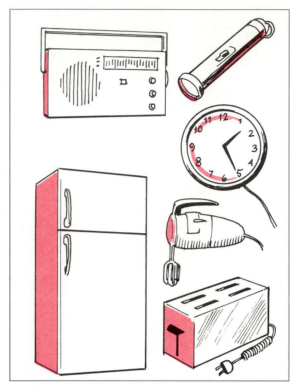

1. Figure C shows some electrical devices. How many can you name?

   Radio  Flashlight _____

   Clock  Hand mixer _____

   Refrigerator  Toaster _____

   _____

2. How many other electrical devices can you name?

   _____

   _____

   _____

   _____

**Figure C**

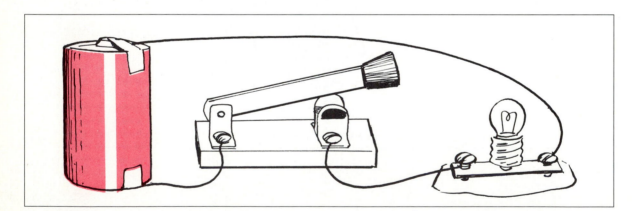

**Figure D**

3. Is this circuit complete or incomplete? _____ Incomplete _____

4. Are electrons moving? _____ No _____

5. Does the bulb light up? _____ No _____

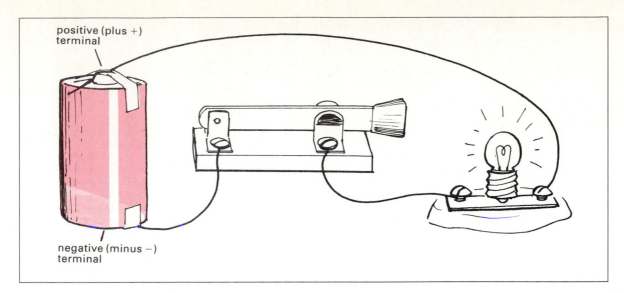

positive (plus +)
terminal

negative (minus −)
terminal

**Figure E**

6. Is this circuit complete or incomplete? _____ Complete _____

7. Are electrons moving? _____ yes _____

8. Does the bulb light up? _____ yes _____

9. Electricity flows from minus to plus. Draw arrows near the wires, the switch, and battery to show this path.

## FILL IN THE BLANK

*Complete each statement using a term or terms from the list below. Write your answers in the spaces provided. Some words may be used more than once.*

| | | |
|---|---|---|
| complete | negative | positive |
| generators | move along a path | circuit |
| incomplete | toaster | do not move along a path |

1. In static electricity, electrons _____ do not move along a path _____ .

2. In current electricity, electrons _____ move along a path _____ .

3. The path along which electrons move is called a _____ circuit _____ .

4. Electrons do not move in an _____ incomplete _____ circuit.

5. Electrons do flow in a _____ complete _____ circuit.

6. Electrons leave a dry cell through the _____ negative _____ terminal.

7. Electrons return to a dry cell through the _____ positive _____ terminal.

8. Large amounts of electricity are made by _____ generators _____ .

9. An example of an electrical appliance is a _____ toaster _____ .

## IDENTIFY THESE ELECTRICAL SYMBOLS

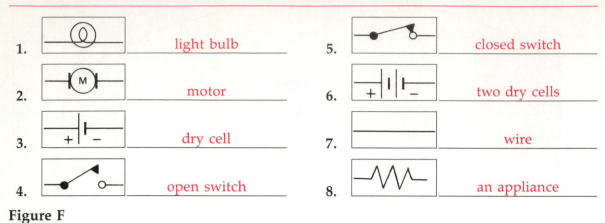

1. light bulb
2. motor
3. dry cell
4. open switch
5. closed switch
6. two dry cells
7. wire
8. an appliance

**Figure F**

## NOW LET'S DRAW!

*Draw these electrical symbols. [But first cover the top of this page.]*

| | | |
|---|---|---|
| **1.** | one dry cell | |
| **2.** | two dry cells connected together | |
| **3.** | wire | |
| **4.** | light bulb | |
| **5.** | motor | |
| **6.** | open switch | |
| **7.** | closed switch | |
| **8.** | an appliance (resistance) | |

126

## MATCHING

*Match each term in Column A with its description in Column B. Write the correct letter in the space provided.*

**Column A**

_d_ 1. flow of electrons

_b_ 2. circuit

_a_ 3. minus terminal

_e_ 4. plus terminal

_c_ 5. light bulb

**Column B**

a) where electrons leave

b) path for moving electron

c) an electrical device

d) electric current

e) where electrons return

## TRUE OR FALSE

*In the space provided, write "true" if the sentence is true. Write "false" if the sentence is false.*

_True_ 1. Electric current is the flow of electrons.

_False_ 2. Static electricity lights our homes.

_True_ 3. Most of our electricity comes from generators.

_False_ 4. The path that electric current follows is called a circus.

_False_ 5. Electrons leave a battery from the plus terminal.

_True_ 6. Electrons return to a battery through the plus terminal.

_False_ 7. The inside of a battery is filled with zinc.

_False_ 8. Batteries give static electricity.

_True_ 9. Generators make electric current.

_True_ 10. Electrons stop moving in an incomplete circuit.

## WORD SCRAMBLE

*Below are several scrambled words you have used in this Lesson. Unscramble the words and write your answers in the spaces provided.*

1. NUTRECR            CURRENT

2. TIRCCUI            CIRCUIT

3. TARBETY            BATTERY

4. REMLATIN            TERMINAL

5. CELOTERN            ELECTRON

## REACHING OUT

*Why don't we get most of our electricity from batteries?*

Batteries alone cannot provide the large amounts of power we need today. Generators are needed to make large amounts of electricity.

**Figure G**

128

# What is a series circuit? 21

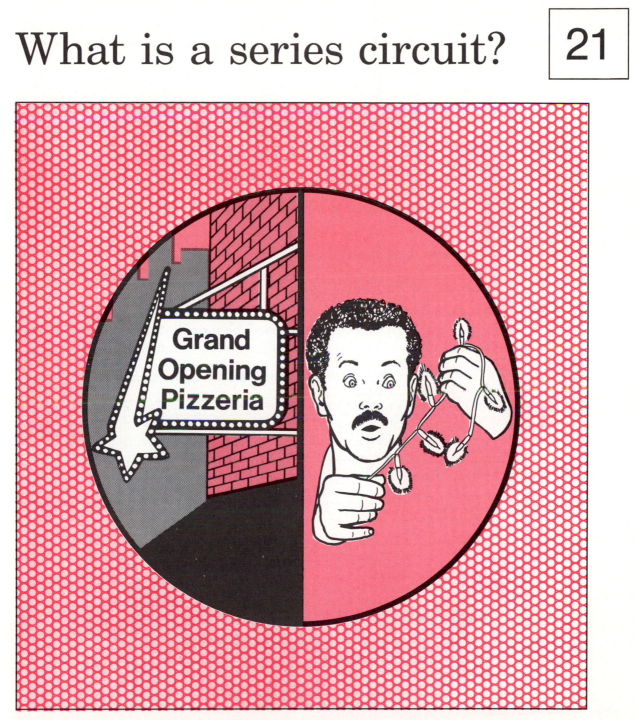

**series circuit:** an electrical hook-up in which the current has only one path

# LESSON 21 | What is a series circuit?

How many light bulbs are there in your home? How many other electrical devices do you have?

Must they all be working if you want to use just one?

Do they all stop working if you shut off just one?

Of course not! Homes are not wired that way. But there are electrical hook-ups that work so that all electrical devices on the circuit are either on or off. This kind of electrical hook-up is called a series circuit.

There are two important things to remember about a series circuit:

1. Electrons have only one path to follow in a series circuit. Each electrical device is connected along this one path. Because of that, the electricity cannot go to just one device. It must move through all. If you turn off any electrical device, you will turn them all off. If you turn that device back on, you will turn on all the devices.

2. The electrical devices, or appliances, share the electrical pressure in a series circuit. If you add electrical appliances, each one gets less electrical pressure. For example, suppose you have light bulbs along a series circuit. Then you add more bulbs. What would happen? Each bulb would give off less light.

Why bother with series circuits? They don't make sense for homes, schools, or factories! But there are special uses for series circuits. Parts of computers, radios, and television sets are wired in series. Parts of space rockets are too!

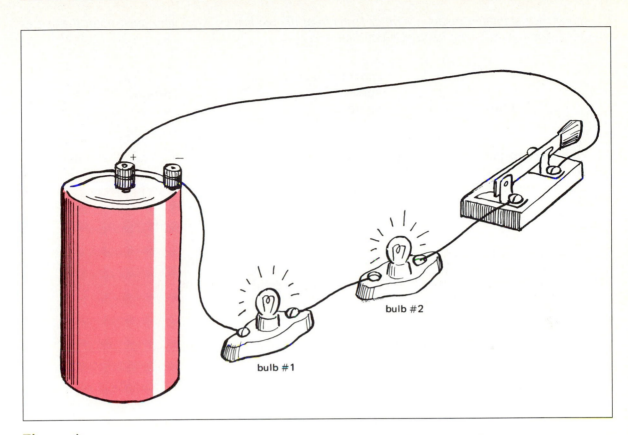

**Figure A**

1. Trace the path of the electrons in this series circuit. (Draw in arrows along the circuit.)

2. In this circuit, the electricity has _____one_____ path(s) to follow.
   <sub>one, two</sub>

3. This circuit is _____complete_____ .
   <sub>complete, incomplete</sub>

4. Where does the electricity have to go before it reaches bulb #2? ___through bulb #1___

5. If bulb #1 were to go out, bulb #2 would _____go out_____ .
   <sub>stay lit, go out</sub>

6. If bulb #2 were to go out, bulb #1 would _____go out_____ .
   <sub>stay lit, go out</sub>

7. In this circuit, each bulb _____is not_____ getting the full electrical pressure.
   <sub>is, is not</sub>

8. If more bulbs were added to this circuit, each bulb would give off

   _____less_____ light.
   <sub>more, less</sub>

9. If this circuit had only one bulb, it would give off _____more_____ light.
   <sub>more, less</sub>

# FILL IN THE BLANK

*Complete each statement using a term or terms from the list below. Write your answers in the spaces provided. Some words may be used more than once.*

| | | |
|---|---|---|
| go off | moving electrons | switched on |
| less | series | are not |
| one | share | |

1. The circuit you are learning about in this lesson is the _____series_____ circuit.

2. In a series circuit, electrons have only _____one_____ path to follow.

3. In a series circuit, when one appliance is shut off, all other appliances _____go off_____ .

4. In a series circuit, when one appliance is switched on, all other appliances must be _____switched on_____ .

5. In a series circuit, the appliances _____share_____ the electrical pressure.

6. In a series circuit, when you add more appliances, each appliance gets _____less_____ power.

7. Homes, factories, and schools _____are not_____ wired in series.

8. Electric current comes from _____moving electrons_____ .

# MATCHING

*Match each term in Column A with its description in Column B. Write the correct letter in the space provided.*

### Column A

__c__ 1. charged atoms that are not moving

__e__ 2. moving electrons

__a__ 3. series circuits

__d__ 4. minus terminal

__b__ 5. plus terminal

### Column B

a) only one path for electrons to move

b) ending point of a circuit

c) static electricity

d) starting point of a circuit

e) electric current

# UNDERSTANDING SERIES CIRCUITS

*Four series circuits are shown below. Use arrows to show the path of the electricity in each one.*

**Figure B**

How many paths are there in this circuit?

one

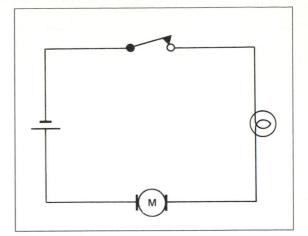

**Figure C**

How many paths are there in this circuit?

one

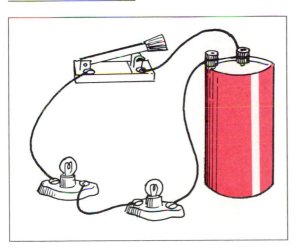

**Figure D**

How many paths are there in this circuit?

one

**Figure E**

How many paths are there in this circuit?

one

# COMPLETE THE CHART

*Use electrical symbols to draw these series circuits.*

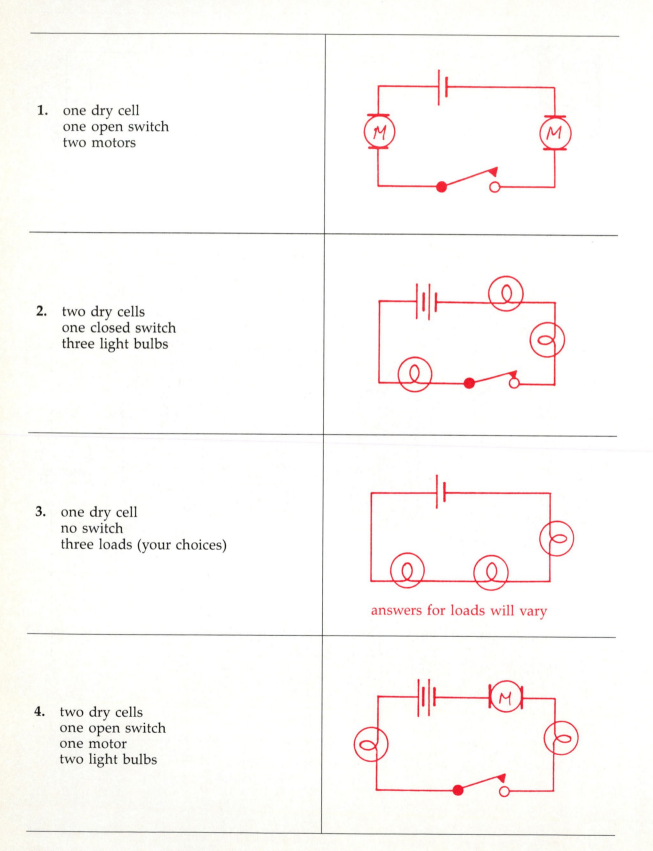

1. one dry cell
   one open switch
   two motors

2. two dry cells
   one closed switch
   three light bulbs

3. one dry cell
   no switch
   three loads (your choices)

   answers for loads will vary

4. two dry cells
   one open switch
   one motor
   two light bulbs

# What is a parallel circuit?

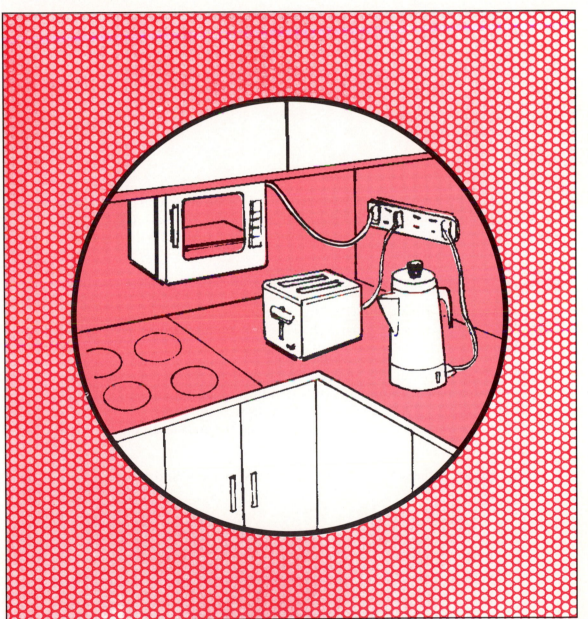

**parallel circuit:**  an electrical hook-up in which the current has more than one path

# LESSON 22 | What is a parallel circuit?

You walk into your home and switch on the TV. You switch on only the TV. You don't have to switch on the toaster and broiler, the hair dryer and all your lights—You don't have to because your home is not wired in series. Your home is wired in parallel.

There are two important facts you should know about **parallel circuits:**

1. In a parallel circuit, the electrons have more than one path to follow. Each appliance has its own path. This lets you use or shut off only one appliance at a time.

2. In a parallel circuit, the appliances do not share the electrical pressure. Each appliance gets the full voltage it needs. Adding more loads does not weaken the force. Each load still works with full power. For example, adding more bulbs to a parallel circuit does not make each bulb give off less light.

Parallel circuits make sense for use in homes, schools, and factories.

# AN EXAMPLE OF PARALLEL CIRCUIT

bulb #2

bulb #1

**Figure A**

*Look at Figure A. Then answer the questions.*

1.  How many bulbs are in this parallel circuit? _____ two _____

2.  How many paths does the electricity have to follow? _____ two _____ Follow the paths that are shown with your pencil.

3.  Is this circuit complete or incomplete? _____ complete _____

4.  Do the bulbs light up? _____ yes _____

5.  Does the electricity have to pass through bulb #1 for bulb #2 to light up? _____ no _____

6.  If bulb #2 were to blow out, bulb #1 would _____ stay lit _____ .
    <br>stay lit, go out

7.  If bulb #1 were to blow out, bulb #2 would _____ stay lit _____ .
    <br>stay lit, go out

8.  If a third bulb were added, bulbs #1 and #2 would
    _____ give off the same amount of light _____ .
    <br>give off less light, give off the same amount of light

9.  The bulbs in this circuit _____ do not _____ share the electrical pressure.
    <br>do, do not

10. Your home is wired _____ in parallel _____ .
    <br>in parallel, in series

137

## COMPLETING SENTENCES

*Choose the correct word or term for each statement. Write your choice in the spaces provided.*

1. Homes, schools and factories, _____are not_____ wired in series.
   *are, are not*

2. This school is wired in _____parallel_____ .
   *parallel, series*

3. In a series circuit, electricity has _____one_____ path to follow.
   *one, more than one*

4. In a parallel circuit, electricity has _____more than one_____ path to follow.
   *one, more than one*

5. In a series circuit, when one bulb goes out, the other bulbs _____go off_____ .
   *stay lit, go off*

6. In a parallel circuit, when one bulb shuts off, the other bulbs _____stay lit_____ .
   *stay lit, go off*

7. An extra bulb is added to a series circuit. The other bulbs now give off
   _____less light_____ .
   *less light, the same amount of light*

8. An extra bulb is added to a parallel circuit. The other bulbs now give off
   _____the same amount of light_____ .
   *less light, the same amount of light*

9. In a parallel circuit, you _____can_____ use or shut off one appliance at a time.
   *can, cannot*

10. In a series circuit, you _____cannot_____ use or shut off one appliance at a time.
    *can, cannot*

## MATCHING

*Match each term in Column A with its description in Column B. Write the correct letter in the space provided.*

| Column A | Column B |
|---|---|
| __c__ 1. parallel circuit | **a)** does not change amount of light each bulb gives |
| __b__ 2. series circuit | **b)** loads work together |
| __a__ 3. another bulb added to a parallel circuit | **c)** loads work one at a time |
| __d__ 4. another bulb added to a series circuit | **d)** does change amount of light each bulb gives |

138

## WORKING WITH CIRCUITS

*Look at each circuit. Then answer the questions next to it.*

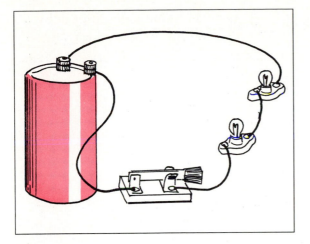

**Figure B**

(Note: Do not count a switch as a load.)

1. What kind of circuit is this?

   <u>series circuit</u>

2. How many paths do the electrons have to follow? <u>one</u>

3. How many loads does this circuit have? <u>two</u>

4. Is the circuit complete or incomplete?

   <u>complete</u>

5. Are the loads working? <u>yes</u>

6. If one bulb were to blow out, the other bulb would <u>shut off</u>
   stay lit, shut off

7. Adding another bulb would make the other two give off

   <u>less light</u> .
   less light, the same amount of light

8. This <u>is not</u> a good way to wire a home.
   is, is not

*Look at Figure C.*

9. What kind of circuit is this?

   <u>parallel</u>

10. How many loads does this circuit have? <u>two</u>

11. How many paths do the electrons have to follow? <u>two</u>

12. Is the circuit complete or incomplete?

    <u>complete</u>

**Figure C**

13. Are the loads working? _____ yes _____

14. If one bulb were to go off, the other bulb would give off

    _____ the same amount of light _____ .
    more light, the same amount of light

15. Adding another bulb would make each bulb give off _____ the same amount of light _____ .
    less light, the same amount of light

16. Is this a good way to wire a home? _____ yes _____

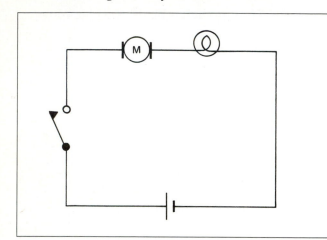

**Figure D**

*Look at Figure D.*

17. What kind of circuit is this? _____ series circuit _____
    parallel, series

18. How many paths do the electrons have to follow? _____ one _____

19. How many loads does this circuit have? _____ two _____

    Name them. _____ motor _____ light bulb _____

20. Is the circuit complete or incomplete? _____ incomplete _____

21. Are the loads working? _____ no _____

22. Is your home wired this way? _____ no _____

140

*Look at Figure E.*

**23.** What kind of circuit is this?

_____parallel_____
parallel, series

**24.** How many paths do the electrons have to follow? _____two_____

**25.** How many loads does this circuit have? _____two_____

Name them. __motor__

light bulb

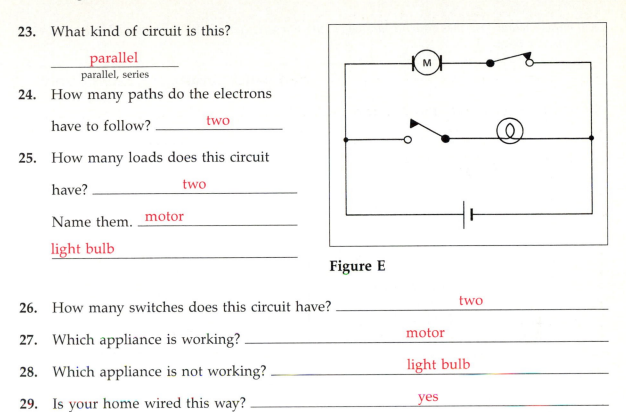

**Figure E**

**26.** How many switches does this circuit have? _____two_____

**27.** Which appliance is working? _____motor_____

**28.** Which appliance is not working? _____light bulb_____

**29.** Is your home wired this way? _____yes_____

*Look at Figure F.*

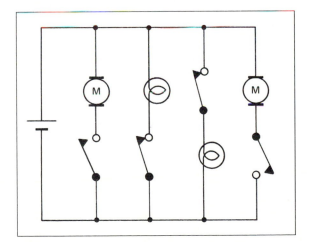

**Figure F**

**30.** What kind of circuit is this?

_____parallel_____

**31.** How many paths do the electrons have to follow? _____four_____

**32.** How many loads does this circuit have? _____four_____

Name them. __two motors__

two light bulbs

**33.** How many switches does this circuit have? _____four_____

**34.** Which loads are working? __two light bulbs__

**35.** Which loads are not working? __two motors__

**36.** Is your school wired this way? __yes__

141

## COMPLETE THE CHART

Each phrase below describes either a parallel circuit or a series circuit. Which one is it? Put a check (✔) in the proper box.

| | Parallel Circuit | Series Circuit |
|---|:---:|:---:|
| **1.** only one path for the electricity to follow | | ✔ |
| **2.** more than one path for the electricity to follow | ✔ | |
| **3.** loads work or shut off one at a time | ✔ | |
| **4.** all loads are on or all loads are off | | ✔ |
| **5.** appliances share the voltage | | ✔ |
| **6.** appliances do not share the voltage | ✔ | |
| **7.** good way to wire a home | ✔ | |
| **8.** not a good way to wire homes | | ✔ |
| **9.** an extra bulb makes the others less bright | | ✔ |
| **10.** an extra bulb does not change the brightness of the others | ✔ | |

## WORD SCRAMBLE

Below are several scrambled words you have used in this Lesson. Unscramble the words and write your answers in the spaces provided.

1. RESSIE            _____series_____

2. LAPELRAL         _____parallel_____

3. TRAGEENOR       _____generator_____

4. SPALNAPCIE      _____appliances_____

5. GRINIW            _____wiring_____

## REVIEWING ELECTRICAL SYMBOLS

*Draw the following electrical symbols.*

| | | |
|---|---|---|
| **1.** | open switch | |
| **2.** | closed switch | |
| **3.** | one dry cell | |
| **4.** | two dry cells | |
| **5.** | wire | |
| **6.** | motor | |
| **7.** | light bulb | |

## TRUE OR FALSE

*In the space provided, write "true" if the sentence is true. Write "false" if the sentence is false.*

| | | |
|---|---|---|
| False | **1.** | A dry cell gives static electricity. |
| False | **2.** | Static electricity lights our homes. |
| True | **3.** | Static electricity causes lightning. |
| False | **4.** | A safe place to stay during a lightning storm is under a tree. |
| True | **5.** | Electricity is useful. |
| True | **6.** | Electricity can be dangerous. |
| True | **7.** | This school is wired in parallel. |
| False | **8.** | Your home is wired in series. |
| True | **9.** | A parallel circuit lets you use or shut off one appliance at a time. |
| False | **10.** | Appliances wired in parallel share the electrical pressure. |

*Draw these circuits. Use electrical symbols.*

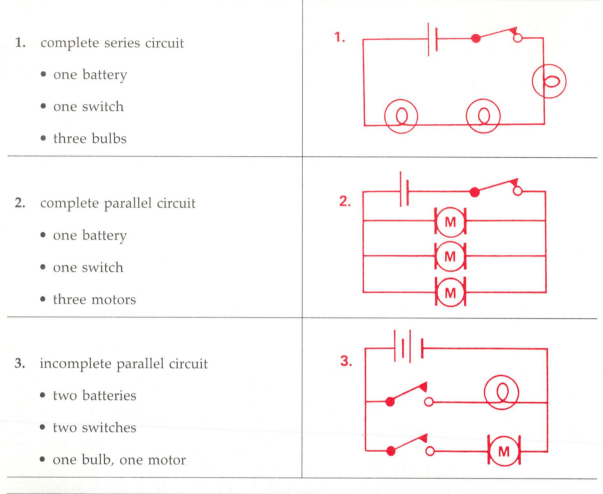

1. complete series circuit
   - one battery
   - one switch
   - three bulbs

2. complete parallel circuit
   - one battery
   - one switch
   - three motors

3. incomplete parallel circuit
   - two batteries
   - two switches
   - one bulb, one motor

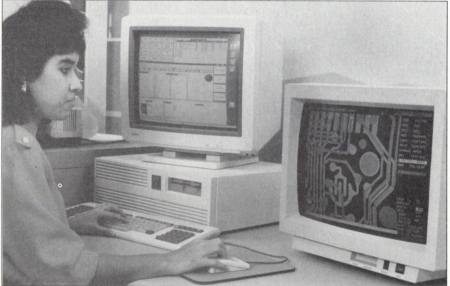

**Figure G**

# What is electrical resistance?

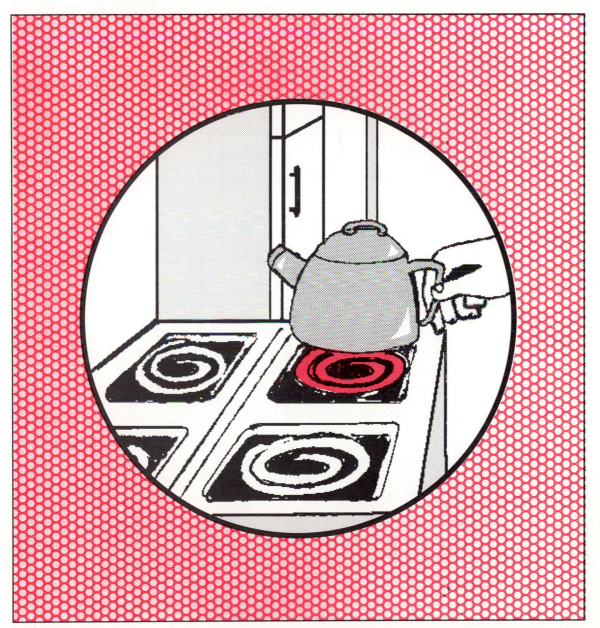

**resistance:**   tendency to slow or stop electric current

# LESSON 23 | What is electrical resistance?

Imagine that you are walking against a strong wind. It isn't easy to walk. The wind is slowing you down. It is trying to stop you. We say the wind resists your movement.

Everything that moves meets some kind of **resistance.** Even electricity meets resistance.

Electric wire resists the flow of electrons. It tries to stop the electrons. The resistance makes the atoms and molecules rub together. This rubbing, or friction, builds heat. The greater the resistance, the greater the heat.

Electrical resistance can be slight—or very great—or in-between. Resistance depends mainly on three things. They are: wire length, wire thickness, and the kind of metal the wire is made of.

**LENGTH OF WIRE** Long wires resist electricity more than short wires do. The longer the wire, the more resistance.

**THICKNESS OF WIRE** Thin wires resist electricity more than thick wires do. The thinner the wire, the greater the resistance.

**KIND OF METAL** Some metals resist electricity more than others. Silver resists electricity the least. Copper resists electricity less than most metals. Metals that offer little resistance are good for electrical wiring. Most electrical wiring is made of copper.

Nichrome [NIE krome] is made of nickel and chromium. Nichrome offers great resistance to electricity. Metals that offer great resistance are good for producing heat. They can be used in toasters and electric irons.

## RESISTANCE AND WIRE THICKNESS

Two wires A and B are shown below. They are the same length.

How are they different? _A is thicker, B is thinner_

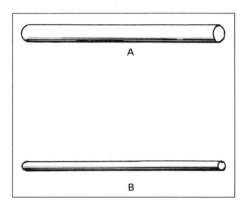

**Figure A**

*Now fill in the blanks below using the letters A and B.*

1. Electrons have more room to move along wire ___A___.

2. Electrons have less room to move along wire ___B___.

3. Electrons rub more along wire ___B___.

4. Electrons rub less along wire ___A___.

5. Which wire resists the electrons more? ___B___

6. Which wire resists the electrons less? ___A___

7. There is more friction along wire ___B___.

8. There is less friction along wire ___A___.

9. Which wire stays cooler? ___A___

10. Which wire becomes warmer? ___B___

## CONCLUSION:

11. Thin wire resists electricity ___more___ than thick wire.
    <br>more, less

## RESISTANCE AND WIRE LENGTH

Two wires C and D are shown below. They are both the same thickness.

How are they different? _Wire C is shorter, wire B is longer_

_____

**Figure B**

_Now fill in the blanks below, using the letters C and D._

1. Electrons move farther along wire _____D_____ .

2. Electrons move a shorter distance along wire _____C_____ .

3. Electrons rub more along wire _____D_____ .

4. Electrons rub less along wire _____C_____ .

5. Which wire resists the electrons more? _____D_____

6. Which wire resists the electrons less? _____C_____

7. There is more friction along wire _____D_____ .

8. There is less friction along wire _____C_____ .

9. Which wire stays cooler? _____C_____

10. Which wire becomes warmer? _____D_____

## CONCLUSION:

Long wire resists electricity _____more_____
than short wire.    more, less

**Figure C**

## COMPLETING SENTENCES

*Choose the correct word or term for each statement. Write your choice in the spaces provided.*

1. To "resist" means to ___try to stop___ .
   <u>help, try to stop</u>

2. Electrical resistance is caused by ___friction___ .
   <u>friction, switches</u>

3. Friction comes from ___rubbing___ .
   <u>wires, rubbing</u>

4. Friction produces ___heat___ .
   <u>electrons, heat</u>

5. More friction means ___more___ heat.
   <u>more, less</u>

6. Less friction means ___less___ heat.
   <u>more, less</u>

7. Long wire resists electricity ___more___ than short wire.
   <u>more, less</u>

8. Thick wire resists electricity ___less___ than thin wire.
   <u>more, less</u>

9. Nichrome is a ___high___ resistance wire.
   <u>high, low</u>

10. Copper is a ___low___ resistance wire.
    <u>high, low</u>

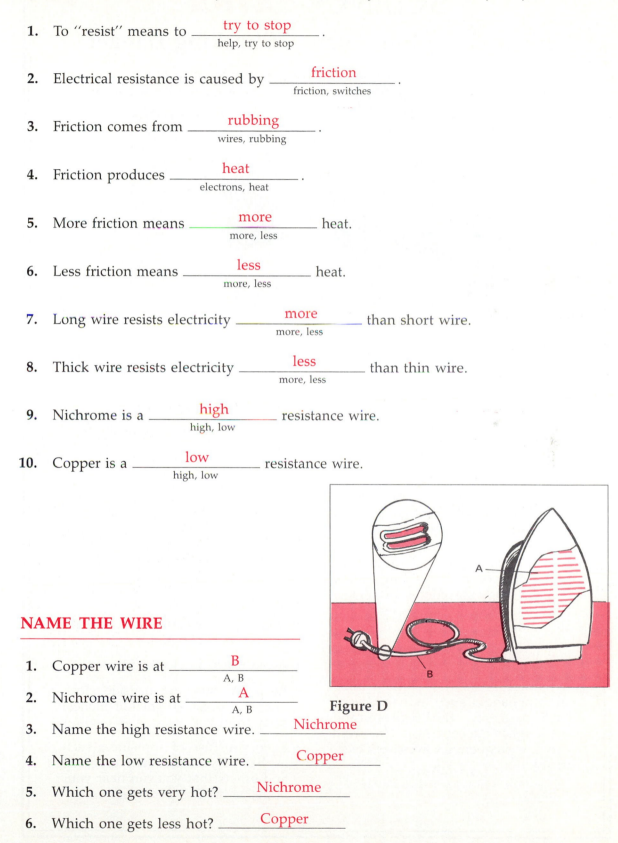

**Figure D**

## NAME THE WIRE

1. Copper wire is at ___B___
   <u>A, B</u>

2. Nichrome wire is at ___A___
   <u>A, B</u>

3. Name the high resistance wire. ___Nichrome___

4. Name the low resistance wire. ___Copper___

5. Which one gets very hot? ___Nichrome___

6. Which one gets less hot? ___Copper___

149

# SCIENCE EXTRA

## Electricity: More Than A Shocking Danger?

At this moment—while you read this—electromagnetic energy is bombarding you. You can't see it or feel it, but it is there. This energy ranges from high frequency x-rays and ultraviolet rays to low frequency radio waves.

We already know that too much exposure to ultraviolet and x-rays can cause cancer. Now, some recent research has led scientists to suspect a link between long-term exposure to very low frequency waves and increased risk of leukemia and other forms of cancer. This is disturbing because everyone comes in contact with low frequency waves just about all the time.

Low frequency waves are given off by anything that is conducting electricity: power lines, household wiring, and appliances such as hair dryers, TV's, toasters . . . the list goes on and on. It doesn't seem possible to avoid very low frequency waves from electricity.

What might this electromagnetic radiation do to human health? Several research projects are already in progress. They are designed to learn if low frequency waves might:

- reduce the body's natural ability to fight cancer

- upset the chemistry of the body's cell membranes

- affect normal brain function

- affect normal functions of endocrine glands, such as the gonads.

Here are some tips you can follow to play it safe:

- Don't spend a lot of time beneath power lines

- Avoid using an electric blanket

- Keep TV sets and bedside clock radios at least 70 cm (30 inches) away from you

- Keep your use of appliances such as hair dryers, vacuum cleaners, and others that you run near your body, as brief as possible.

# What are amperes, volts, and ohms?

24

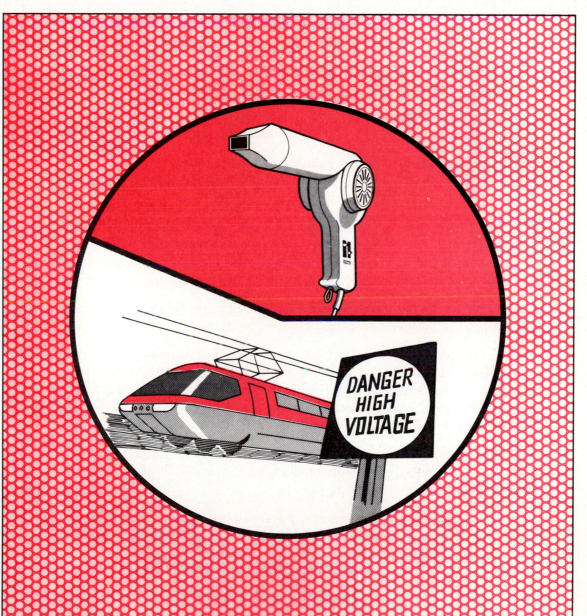

**ampere:**   unit for measuring the number of electrons moving past a point in a circuit
**electromotive force (EMF):**   electrical pressure
**volt:**   unit for measuring electrical pressure
**ohm:**   unit of electrical resistance

# LESSON 24 | What are amperes, volts, and ohms?

How do you measure temperature? In degrees. You measure time in minutes, hours, days, etc. How do you measure length? Weight?

We use different units to measure different things. There are special units to measure electricity too. Three of the most important are **ampere, volt,** and **ohm.**

**AMPERES** [AM peers] The size of an electric current depends on how many electrons pass a point in a circuit every second. The greater the number of electrons, the larger the current. Fewer electrons mean a smaller current.

The size of an electric current is measured in amperes [amps]. We can say that ampere is another name for electric current.

**VOLTS** Nothing moves by itself. A force is needed to make something move. Electricity needs a force to move it. Electrons move in a circuit because a force pushes them. The name for the force or pressure that pushes electrons is **electromotive force.** It is often called EMF. The strength of the EMF is measured in volts.

**OHMS** [OMES] Ohms measure the resistance to the flow of electrons. You know that a wire resists the flow of electrons. The amount of resistance is measured in ohms.

There is a connection between amps, volts, and ohms. When one changes, there must be a change in one or both of the others. There is a rule for figuring these changes. It is called Ohm's Law.

**Figure A**

| VOLTS | AMPERES (AMPS) | OHMS |
|---|---|---|
| The force that moves electrons in a circuit | The number of electrons that are moving | Resistance—the force that tries to stop or slow the electrons |

**Figure B**

This number of electrons passing a point in a wire every second is one ampere of current. Which one is easier to say—one ampere or 6,281,000,000,000,000,000 electrons?

Different electrical devices use different amperes.

- A 100-watt light bulb uses about 1 ampere.

- An electric iron or broiler uses about 10 to 12 amperes.

**Figure C**

# FILL-IN QUESTIONS

*Fill in the correct answer for each of the following.*

1. Another name for electric current is _____amperes_____ .

2. Amperes tell us how many _____electrons_____ move past a point in a circuit every second.

3. EMF stands for _____electromotive force_____ .

4. Electrical force or pressure is measured in units called _____volts_____ .

5. Electrical resistance is measured in units called _____ohms_____ .


# YOUR OWN WORDS, PLEASE!

*Use your own words to tell what each of the following is.*

1. EMF _____pressure that pushes electrons_____

2. VOLTS _____units of measure for EMF_____

3. AMPERES _____units that measure amount of electric current_____

4. OHMS _____units that measure electrical resistance_____


# MATCHING

*Match each term in Column A with its description in Column B. Write the correct letter in the space provided.*

| Column A | | Column B |
|---|---|---|
| __d__ 1. volts | | a) electrical resistance |
| __e__ 2. amps | | b) path for moving electrons |
| __b__ 3. circuit | | c) relationship between volts, amps, and ohms |
| __c__ 4. Ohm's Law | | d) electrical pressure |
| __a__ 5. ohms | | e) number of electrons passing a point in a wire |

## TRUE OR FALSE

*In the space provided, write "true" if the sentence is true. Write "false" if the sentence is false.*

__False__ 1. EMF stands for a number of electrons.

__False__ 2. Another name for resistance is ampere.

__True__ 3. Volts measure electrical pressure or force.

__True__ 4. Different circuits have different amps, volts, and ohms.

__False__ 5. If volts change, then amps and ohms stay the same.

## WORD SCRAMBLE

*Below are several scrambled words you have used in this Lesson. Unscramble the words and write your answers in the spaces provided.*

1. PREAME _____ampere_____

2. SEPRURES _____pressure_____

3. MOH _____ohm_____

4. TOLV _____volt_____

5. TRENRUC _____current_____

## COMPLETING SENTENCES

*Choose the correct word or term for each statement. Write your choice in the spaces provided.*

1. Electricity that is not moving is called ___static___ electricity.
   <br>static, current

2. Electricity that is moving is called ___current___ electricity.
   <br>static, current

3. Friction produces ___static___ electricity.
   <br>static, current

4. A dry cell produces ___current___ electricity.
   <br>static, current

5. The electricity we use is ___current___ electricity.
   <br>static, current

6. Current electricity is the flow of ___electrons___.
   <br>atoms, electrons

7. Most current electricity comes from ___generators___.
   <br>generators, dry cells

8. The circuit that has only one path to follow is the _____series_____ circuit.
   <sub>parallel, series</sub>

9. The circuit that has more than one path to follow is the _____parallel_____ circuit.
   <sub>parallel, series</sub>

10. Homes, schools, and factories are wired in _____parallel_____.
    <sub>parallel, series</sub>

11. "All appliances on or all off" tells us that the circuit is wired in _____series_____.
    <sub>parallel, series</sub>

12. "Any number of appliances on or off" tells us that the circuit is wired in

    _____parallel_____.
    <sub>parallel, series</sub>

13. Change of wire length or thickness _____does_____ change electrical resistance.
    <sub>does, does not</sub>

14. Nichrome is a _____high_____ resistance wire.
    <sub>high, low</sub>

15. A high resistance wire builds _____much_____ heat.
    <sub>little, much</sub>

16. A low resistance wire builds _____little_____ heat.
    <sub>little, much</sub>

17. Copper is a _____low_____ resistance wire.
    <sub>high, low</sub>

18. The size of an electric current is measured in _____amperes_____.
    <sub>amperes, volts, ohms</sub>

19. Electrical pressure is measured in _____volts_____.
    <sub>amperes, volts, ohms</sub>

20. Electrical resistance is measured in _____ohms_____.
    <sub>amperes, volts, ohms</sub>

## REACHING OUT

*Do some research to answer this question.*

The distance from the east coast to the west coast of the United States is about 7770 kilometers (about 3000 miles). How many times can electricity travel this distance in just one second?

In one second, electricity travels 300,000 kilometers. It can travel across the U.S. 39 times

in one second.

# What are magnets?

**magnet:** a metal that can attract certain other metals
**lodestone:** a rock that is a magnet
**magnetite:** another name for lodestone
**alloy:** two or more metals melted together

# LESSON 25 | What are magnets?

Did you ever use magnets? If you hold two magnets close to each other, they pull together. Turn one around, and they push apart. You can actually feel the invisible force they give off.

Magnets are vital in industry. They can be helpful around the house too. Magnets hold notes on a refrigerator door—and pick up pins or tacks. A penny won't stick to a refrigerator door. It won't pick up pins or tacks. A magnet will.

What is a **magnet?** Why does it act the way it does? Why don't other things act like magnets?

Only certain materials can become magnets. A material that can become a magnet is called a magnetic substance. A magnetic substance can also be picked up by a magnet.

There are only three good magnetic substances. They are the metals iron, nickel, and cobalt. Iron is the most magnetic.

One kind of magnet is found in nature. The rock **magnetite** [MAG nuh tite] is a natural magnet. It has bits of iron in it. Magnetite is also called **lodestone.** Most magnets are not natural. They are made by people. Such magnets are called artificial magnets.

Most artificial magnets are made of **alloys** [AL oiz]. An **alloy** is a mixture of metals melted together. Steel is an important alloy. Most magnets are made of steel. Two other alloys are used to make extra-strong magnets. These alloys are alnico [AL nick oe] and Permalloy [PUR muh loy]. Both alloys contain one or more of the three magnetic substances.

Magnets are made in many sizes, shapes, and strengths. They are used in telephones, in motor vehicles, and in many other everyday uses.

# MAGNETS

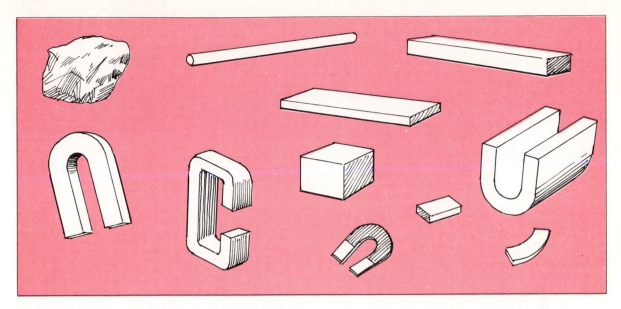

**Figure A**

Magnets come in many sizes and shapes.

Magnetic iron (lodestone) was discovered about 2,000 years ago in a part of Asia called Magnesia. The terms <u>magnet</u> and <u>magnetism</u> come from this place.

## FILL IN THE BLANK

*Complete each statement using a term or terms from the list below. Write your answers in the spaces provided. Some words may be used more than once.*

| | | |
|---|---|---|
| iron | magnetite | Permalloy |
| alloy | picked up | nickel |
| artificial | people | natural |
| magnet | cobalt | lodestone |
| alnico | | |

1. A magnetic substance is a substance that can be _____picked up_____ by a magnet.

2. A magnetic substance can also be made into a _____magnet_____.

3. The three good magnetic substances are: _____iron_____, _____nickel_____,

   and _____cobalt_____.

4. The metal found in most magnets is _____iron_____.

5. Magnets that are found in nature are called _____natural_____ magnets.

6. The name of a natural magnet is _____magnetite_____ . It is also called

   _____lodestone_____ .

7. The opposite of natural is _____artificial_____ .

8. Most magnets are made by _____people_____ .

9. A mixture of metals melted together is called an _____alloy_____ .

10. Very powerful magnets are made of the alloys _____alnico_____ and

    _____Permalloy_____ .

## TESTING FOR MAGNETIC MATERIALS

Touch each material listed below with a magnet. See if it is magnetic or nonmagnetic. Put a check [✔] in the proper box.

The last two lines have been left blank. Select two materials not on the list. Test them and write the results on lines 9 and 10.

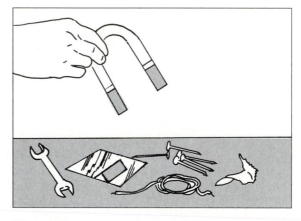

Figure B

| | Material | Magnetic | Nonmagnetic |
|---|---|---|---|
| **1.** | glass | | ✔ |
| **2.** | iron nail | ✔ | |
| **3.** | paper | | ✔ |
| **4.** | plastic | | ✔ |
| **5.** | nickel (not a coin) | ✔ | |
| **6.** | copper | | ✔ |
| **7.** | cobalt | ✔ | |
| **8.** | steel (not stainless) | ✔ | |
| **9.** | Answers will vary | | |
| **10.** | Answers will vary | | |

## TRUE OR FALSE

*In the space provided, write "true" if the sentence is true. Write "false" if the sentence is false.*

False     **1.** A magnet is never found in nature.

False     **2.** All magnets are found in nature.

True     **3.** Lodestone is a natural magnet.

True     **4.** Iron is a magnetic substance.

False     **5.** Iron is an alloy.

True     **6.** Steel is an alloy.

True     **7.** Most magnets are made of steel.

False     **8.** The strongest magnets are made of steel.

False     **9.** Copper is a magnetic substance.

False     **10.** Most substances are magnetic.

## MATCHING

*Match each term in Column A with its description in Column B. Write the correct letter in the space provided.*

| | Column A | | Column B |
|---|---|---|---|
| c | **1.** iron, nickel, cobalt | **a)** | mixture of metals |
| e | **2.** magnetite | **b)** | means "not natural" |
| a | **3.** alloy | **c)** | good magnetic metals |
| b | **4.** artificial | **d)** | alloys used for extra-strong artificial magnets |
| d | **5.** alnico and Permalloy | **e)** | natural magnet |

# WORD SCRAMBLE

Below are several scrambled words you have used in this Lesson. Unscramble the words and write your answers in the spaces provided.

1. BLATCO                            <u>cobalt</u>

2. GETMAN   .                        <u>magnet</u>

3. LETNODEOS                         <u>lodestone</u>

4. YOLAL                             <u>alloy</u>

5. CAINOL                            <u>alnico</u>

# REACHING OUT

Try this at home:

Use a magnet to find out which home appliances have magnetic metals or alloys in them. Make a list of those that do.

# How do magnets behave?  26

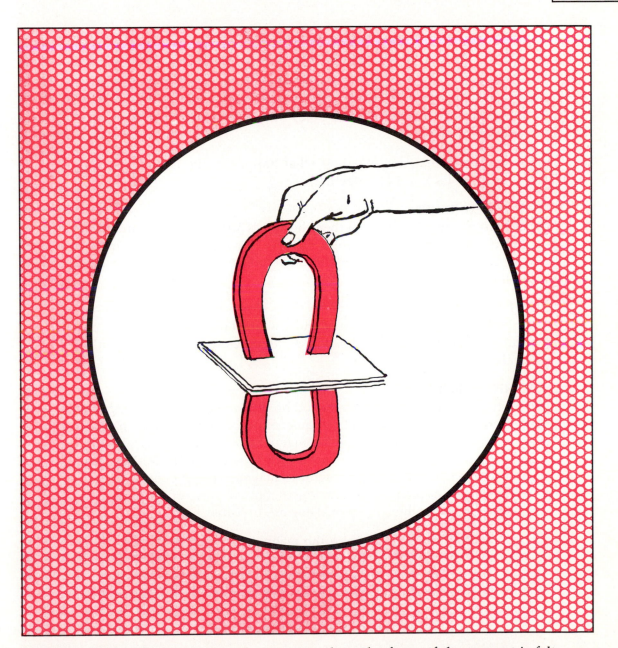

**magnetic field:**   the space around a magnet where the force of the magnet is felt
**magnetic force:**   the push or pull of a magnet upon a magnetic object

# LESSON 26 | How do magnets behave?

Every magnet has two poles—a north pole and a south pole. No matter what shape a magnet comes in, it has two poles, no more and no less. By seeing what you can do with the poles of a magnet, you can learn how magnets behave.

If you hold two magnets together, what happens? The answer depends upon which poles you hold together.

## LIKE POLES REPEL
If you hold two north poles together, they push apart. The same thing will happen if you hold two south poles together. We say that two north poles are like poles. Two south poles are also like poles. Like poles always repel each other.

## UNLIKE POLES ATTRACT
What happens if you hold a north and a south pole together? They pull toward each other. A north pole and a south pole are unlike poles. Unlike poles always attract each other.

## MAGNETIC FORCE
The push or pull that you feel when you hold two magnetic poles together is **magnetic force.** Every magnet can push or pull other magnetic material. The magnetic force is strongest at the magnet's poles. The space around a magnet where magnetic forces can act is called the **magnetic field.** Lines of magnetic force reach through space from a magnet's north pole to its south pole. These lines of force are closest together at the poles of a magnet. You cannot see magnetic lines of force. They are invisible.

A magnetic field is strong close to the magnet. The magnetic field grows weaker the farther you get from the magnet. A magnet can push or pull a magnetic material that is in its magnetic field. A magnet can do this without touching the other object. This ability is why we say that magnets have magnetic energy.

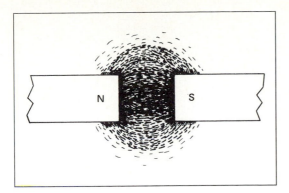

**Figure A** *Magnetic field between unlike poles.*

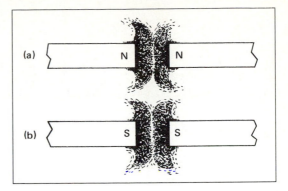

**Figure B** *Magnetic field between like poles.*

*Look at each picture. Then answer the questions.*

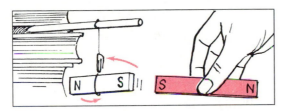

**Figure C**

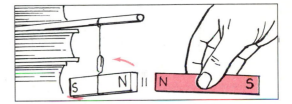

**Figure D**

1. The hand in Figure C is holding the south pole close to the

   _____south_____ pole of the
   north, south

   hanging magnet.

2. They are _____like_____ poles.
   like, unlike

3. The poles _____repel_____.
   attract, repel

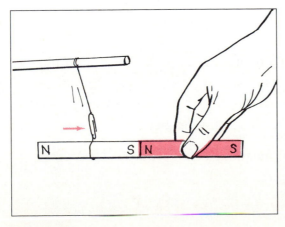

**Figure E**

4. The hand in Figure D is holding the north pole close to the

   _____north_____ pole of the
   north, south

   hanging magnet.

5. They are _____like_____ poles.
   like, unlike

6. The poles _____repel_____.
   attract, repel

7. CONCLUSION: Like poles _____repel_____.
   attract, repel

8. The hand in Figure E is holding the north pole close to the

   _____south_____ pole of the
   north, south

   hanging magnet.

9. They are _____unlike_____ poles.
   like, unlike

10. The poles _____attract_____.
    attract, repel

11. CONCLUSION: Unlike poles

    _____attract_____.
    attract, repel

# HOW TO MAKE A MAP OF A MAGNETIC FIELD

### What You Need (Materials)

bar magnet
thin sheet of paper
iron filings

### How To Do The Experiment (Procedure)

1. Place the magnet on a table.

2. Cover the magnet with the paper.

3. Gently sprinkle the iron filings on the paper.

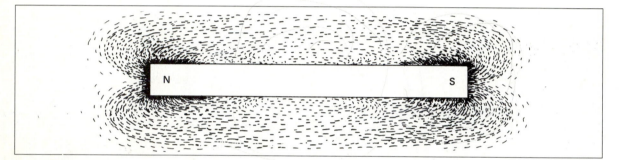

**Figure F**

### What You Learned (Observations)

*Your map should look something like this. Study the map. Then answer these questions.*

1. Most of the iron filings are at ___the poles___ .
   <sub>the poles, the middle</sub>

2. There are fewer iron filings at ___the middle___ .
   <sub>the poles, the middle</sub>

3. A magnet is strongest at ___the poles___ .
   <sub>the poles, the middle</sub>

4. A magnet is weakest at ___the middle___ .
   <sub>the poles, the middle</sub>

5. Most iron filings are ___close to___ the magnet.
   <sub>close to, far from</sub>

6. As you move away from the magnet, there are ___fewer___ iron filings.
   <sub>more, fewer</sub>

### Something To Think About (Conclusions)

1. A magnetic field is strongest ___close to___ a magnet.
   <sub>close to, far from</sub>

2. As you move away from a magnet, the magnetic field becomes ___weaker___ .
   <sub>stronger, weaker</sub>

166

# WORKING WITH MAGNETIC FIELDS

The diagram below shows a bar magnet and its magnetic field. A, B, C, and D are pieces of iron.

*Study the diagram. Then answer the questions.*

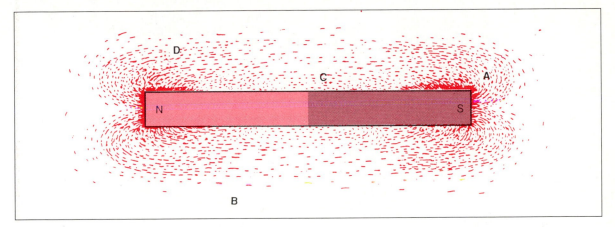

**Figure G**

1. Which pieces of iron are outside the magnetic field? _____ B

2. Which pieces of iron are inside the magnetic field? _____ A C D

3. Look at the pieces that are inside the magnetic field.

   a) Which one does the magnet attract the most? _____ A

   b) Which one does the magnet attract the least? _____ C

# EXPERIMENT WITH MAGNETISM

**Aim:** To find out which substances let magnetic energy pass through them and which substances do not.

## What You Need (Materials)

**Figure H**

magnet
stand with clamp
thin string
steel paper clip
thin pieces of materials listed on the next
  page

## How To Do The Experiment (Procedure)

*Set up the materials as shown on page 167. Then, one at a time, hold the materials listed below between the clip and the magnet. Notice what happens. Does the paper clip drop?*

*Fill in the chart.*

| Material | Does the paper clip drop? [Yes or No] |
|---|---|
| 1. paper | No |
| 2. cloth | No |
| 3. iron | Yes |
| 4. cobalt | Yes |
| 5. glass | No |
| 6. plastic | No |
| 7. nickel | Yes |

## What You Learned (Observations)

1. Which materials did not make the paper clip drop? __paper cloth glass plastic__

2. a) Paper, cloth, glass, and plastic are __non-magnetic__ substances.
   <br>magnetic, non-magnetic

   b) Magnetic energy __does__ go through these substances.
   <br>does, does not

3. Magnetic energy __does__ pass through non-magnetic substances.
   <br>does, does not

4. Which materials did make the paper clip drop? __iron cobalt nickel__

5. a) Iron, nickel, and cobalt are __magnetic__ substances.
   <br>magnetic, non-magnetic

   b) Magnetic energy __does not__ go through these substances.
   <br>does, does not

6. Magnetic energy __does not__ go through magnetic substances.
   <br>does, does not

## Something To Think About (Conclusion)

1. __Non-magnetic__ substances do not affect a magnetic field.
   <br>Magnetic, Non-magnetic

2. __Magnetic__ substances do affect a magnetic field.
   <br>Magnetic, Non-magnetic

# FILL IN THE BLANK

*Complete each statement using a term or terms from the list below. Write your answers in the spaces provided. Some words may be used more than once.*

unlike poles    south pole
like poles     poles
repel       attract
magnetic field    north pole

1. A magnet is strongest at the _____poles_____.

2. One end of a magnet is called the _____north pole_____; the other end is called the _____south pole_____.

3. A south pole and south pole, or a north pole and north pole are called _____like poles_____.

4. A north pole and south pole are called _____unlike poles_____.

5. Like poles _____repel_____.

6. Unlike poles _____attract_____.

7. Two north poles or two south poles will _____repel_____.

8. A north pole and a south pole will _____attract_____.

9. The space around a magnet where the magnetic forces act is called its _____magnetic field_____.

# MATCHING

*Match each term in Column A with its description in Column B. Write the correct letter in the space provided.*

| Column A | | Column B | |
|---|---|---|---|
| __c__ | 1. like poles | a) | make up magnetic field |
| __d__ | 2. unlike poles | b) | magnetic substances |
| __a__ | 3. lines of force | c) | repel |
| __e__ | 4. center of a magnet | d) | attract |
| __b__ | 5. iron, nickel, cobalt | e) | weakest part |

# TRUE OR FALSE

*In the space provided, write "true" if the sentence is true. Write "false" if the sentence is false.*

| | | |
|---|---|---|
| True | 1. | A north pole and a north pole are like poles. |
| False | 2. | Two north poles are the only like poles. |
| False | 3. | Like poles attract. |
| True | 4. | A north pole and a south pole are unlike poles. |
| False | 5. | Unlike poles repel. |
| True | 6. | Lines of force are invisible. |
| False | 7. | A magnet is strongest at the middle. |
| True | 8. | Glass and paper let magnetic energy pass through them. |
| False | 9. | Glass and paper are magnetic substances. |
| False | 10. | Iron lets magnetic energy pass through it. |
| True | 11. | Iron is a magnetic substance. |

# Why are some substances magnetic? 27

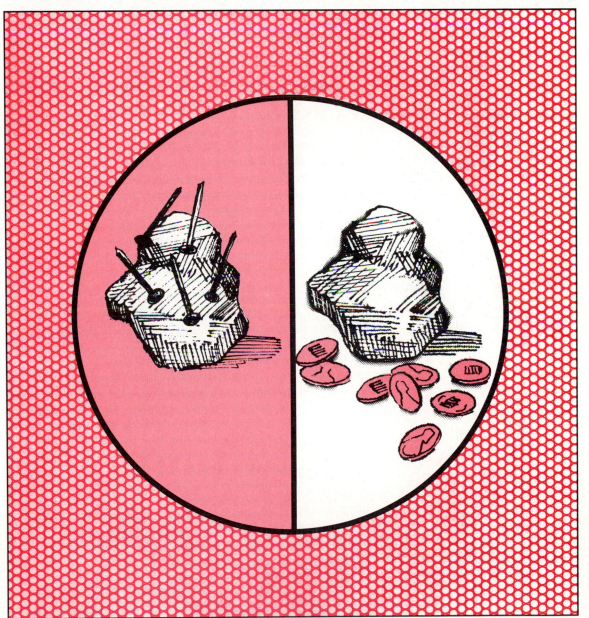

**domain:** a group of lined-up atoms

# LESSON 27 | Why are some substances magnetic?

Imagine a team of soldiers lined up for a drill, all facing one direction. That's how the smallest parts of a magnet act. These parts, called domains, are like tiny magnets. A **domain** [do MAIN] is a group of <u>atoms</u> that are lined up together so that they act as a magnet. Atoms, you may remember, are the smallest parts of any substance. Each atom has a tiny magnetic field.

## WHY IS A DOMAIN LIKE A TINY MAGNET?

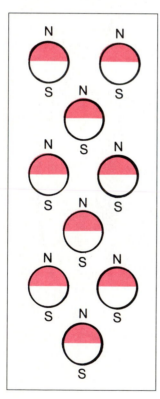

It is because the atoms in each domain line up in the same way. The magnetic field of each atom is lined up in the same direction as all the others. So a domain has a north pole, a south pole, and its own magnetic field.

In a substance that can be made into a magnet, there are many domains. When the substance is not a magnet, the domains are not lined up in one direction. Each domain can face in any direction. In this arrangement, the magnetic forces of all the domains cancel one another out.

When a substance becomes a magnet, the domains all line up in the same direction. That means that all the domains' north poles face one way, and all their south poles face the other way. Now their magnetic fields work together, not against one another.

## BREAKING A MAGNET

<u>A magnet has millions of domains.</u> If a magnet breaks, each piece still has domains. Each broken piece is still a magnet with a north and a south pole. So breaking a magnet makes even more magnets.

Only atoms of iron, nickel, and cobalt can form domains. They are the only magnetic substances. All magnetic alloys contain iron, nickel, or cobalt.

## WHAT DO THE PICTURES SHOW?

Look at Figures A and B. Then answer the questions.

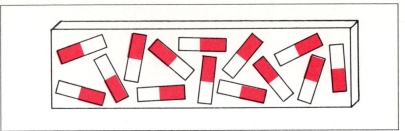

**Figure A** *Single domains in a piece of iron that is _not_ a magnet.*

1. The poles face _____in many directions_____ .
   in one direction, in many directions

2. The magnetic forces _____work against one another_____ .
   work together, work against one another

3. The iron's magnetic forces _____are not_____ felt.
   are, are not

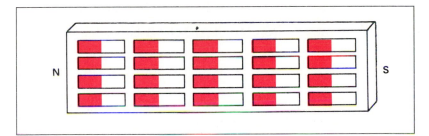

**Figure B** *Groups of atoms in a piece of iron that _is_ a magnet.*

4. What are these groups of atoms called? _____domains_____

5. Their poles face _____in only one direction_____ .
   in only one direction, in many directions

6. The magnetic forces _____work together_____ .
   work together, work against each other

7. The iron's magnetic forces _____are_____ felt.
   are, are not

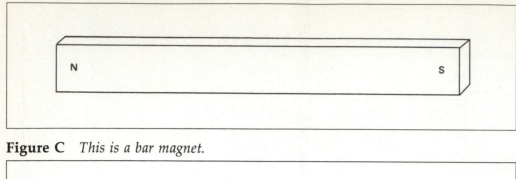

**Figure C**  *This is a bar magnet.*

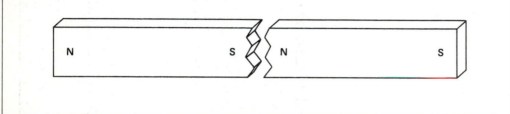

**Figure D**  *This is the same bar magnet broken in half.*

8. Does breaking a magnet destroy the magnet? _____no_____

9. How would four pieces look? Draw the picture in the box below. Label the poles.

**Figure E**

10. How many magnets do you have now? _____four_____

11. What is the smallest part of any magnet? _____a domain_____

174

# FILL IN THE BLANK

Complete each statement using a term or terms from the list below. Write your answers in the spaces provided. Some words may be used more than once.

work together       domains       one domain
magnet             in all directions     ten
one direction

1. Every atom of a magnetic substance is like a tiny __magnet__ .

2. The atoms of matter that is not a magnet face __in all directions__ .

3. The atoms of a magnet form groups called __domains__ .

4. The poles of domains line up in __one direction__ .

5. The magnetic powers of domains __work together__ .

6. The smallest part of a magnet is __one domain__ .

7. If you break a magnet into ten pieces, you end up with __ten__ magnets.

# TRUE OR FALSE

In the space provided, write "true" if the sentence is true. Write "false" if the sentence is false.

__False__ 1. Wood is a magnetic substance.

__True__ 2. Iron is a magnetic substance.

__False__ 3. Every piece of iron has domains.

__True__ 4. Magnetized iron has domains.

__True__ 5. Only magnets have domains.

__False__ 6. Domains work against each other.

__True__ 7. A domain is larger than an atom.

__True__ 8. A magnet has only two poles.

__False__ 9. A magnet can have two north poles.

__False__ 10. You can destroy a magnet by breaking it.

## COMPLETING SENTENCES

*Choose the correct word or term for each statement. Write your choice in the spaces provided.*

1. Atoms normally _____are not_____ lined up.
   <u>are, are not</u>

2. You have a piece of iron that is not a magnet. The atoms _____are not_____ lined up.
   <u>are, are not</u>

3. The atoms of iron _____can_____ be made to line up.
   <u>can, cannot</u>

4. A substance with lined-up atoms is called _____a magnet_____ .
   <u>an alloy, a magnet</u>

5. A group of lined-up atoms is called a _____domain_____ .
   <u>magnetic field, domain</u>

6. Substances like wood, glass, and plastic _____do not_____ form domains.
   <u>do, do not</u>

7. Iron, nickel, and cobalt _____do_____ form domains.
   <u>do, do not</u>

## WORD SCRAMBLE

*Below are several scrambled words you have used in this Lesson. Unscramble the words and write your answers in the spaces provided.*

1. MOANID _____domain_____

2. LICKEN _____nickel_____

3. MOAT _____atom_____

4. POURG _____group_____

5. RONI _____iron_____

# What are temporary and permanent magnets?

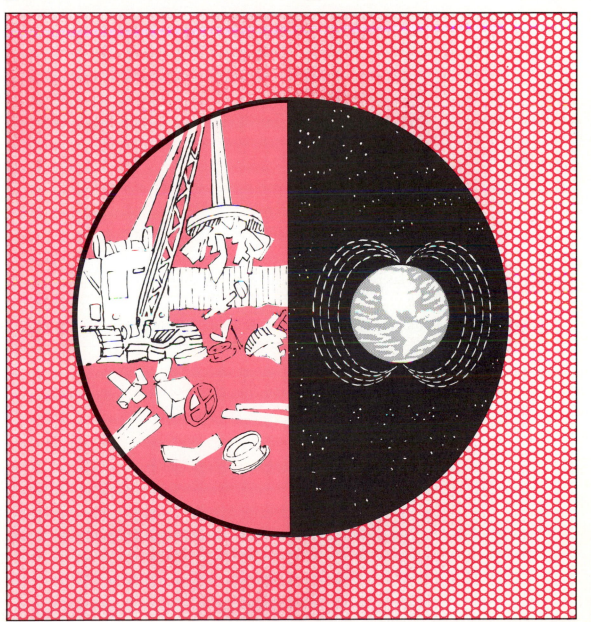

**soft iron:**   iron that loses its magnetism easily.

# LESSON 28 | What are temporary and permanent magnets?

You probably know the word <u>temporary</u> [TEM puh reh ree]. Maybe you know someone who had a temporary job. It lasted for a short time—a week, a month, a summer. The opposite of temporary is <u>permanent</u> [PUR muh nent]. A permanent job lasts for a very long time—years and years.

Something that lasts a short time is temporary. Something that lasts a very long time is permanent. Things that are permanent seem to last forever.

Some magnets are temporary magnets. Others are permanent magnets.

## TEMPORARY MAGNETS
Temporary magnets keep their magnetism for only a short time. Then they lose the magnetism.

Most temporary magnets are made of **soft iron.** Soft iron becomes a magnet very easily. But soft iron loses its magnetism easily.

## PERMANENT MAGNETS
Permanent magnets keep their magnetism. They can be used over and over again. The magnets you use in class are permanent magnets. So are the magnets in your home.

Most permanent magnets are made of steel. Steel does not become a magnet as easily as iron does. But once steel becomes a magnet, it keeps its magnetism.

Certain alloys make extra strong magnets. <u>Alnico</u>, for example, makes extra strong <u>permanent</u> magnets. <u>Permalloy</u> makes extra strong temporary magnets. Alnico is made of iron, aluminum, nickel, and cobalt. Permalloy is made of iron and nickel.

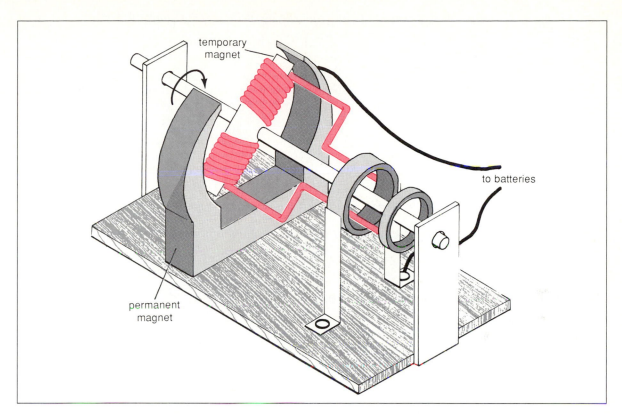

**Figure A** *A motor has temporary and permanent magnets.*

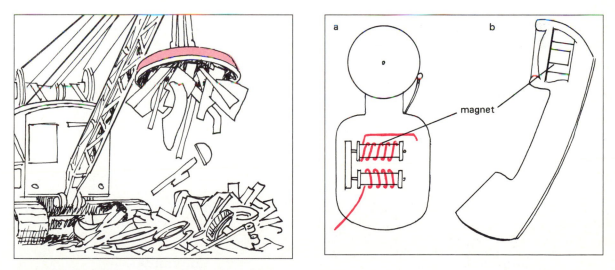

**Figures B and C** *Some temporary magnets are electromagnets.*

Electromagnets have many uses. For example, large electromagnets are used to lift heavy pieces of iron. Electromagnets are also needed in telephones and bells.

# FILL IN THE BLANK

*Complete each statement using a term or terms from the list below. Write your answers in the spaces provided. Some words may be used more than once.*

| | | |
|---|---|---|
| permanent | Permalloy | alloy |
| temporary | steel | lasting a short time |
| magnetism | soft iron | easily |
| loses | alnico | lasting a long time |

1. Temporary means *lasting a short time* .

2. Permanent means *lasting a long time* .

3. A magnet is permanent or temporary depending how long it keeps its *magnetism* .

4. Magnets that keep their magnetism for a short time are *temporary* magnets.

5. Magnets that keep their magnetism for a long time are *permanent* .

6. Temporary magnets are made of *soft iron* .

7. Soft iron becomes a magnet *easily* . Soft iron also *loses* its magnetism easily.

8. Most permanent magnets are made of *steel* .

9. Steel is an *alloy* of iron.

10. Extra strong permanent magnets are made of the alloy *alnico* .

11. Extra strong temporary magnets are made of the alloy *Permalloy* .

# MATCHING

*Match each term in Column A with its description in Column B. Write the correct letter in the space provided.*

| | Column A | | Column B |
|---|---|---|---|
| d | 1. permanent | a) | used for most permanent magnets |
| c | 2. temporary | b) | become extra strong magnets |
| e | 3. soft iron | c) | lasting a short time |
| a | 4. steel | d) | lasting a long time |
| b | 5. alnico and Permalloy | e) | used to make temporary magnets |

## TRUE OR FALSE

*In the space provided, write "true" if the sentence is true. Write "false" if the sentence is false.*

| | | |
|---|---|---|
| False | 1. | All magnets have the same power. |
| False | 2. | All magnets keep their magnetism. |
| True | 3. | Soft iron becomes a magnet easily. |
| True | 4. | Soft iron loses its magnetism easily. |
| False | 5. | Steel becomes a magnet easily. |
| False | 6. | Steel loses its magnetism easily. |
| False | 7. | Soft iron magnets are permanent magnets. |
| True | 8. | Steel magnets are permanent magnets. |
| False | 9. | Alnico magnets are temporary magnets. |
| False | 10. | Every alloy contains iron. |

## MULTIPLE CHOICE

*In the space provided, write the letter of the phrase that best completes each statement.*

b   1. A temporary magnet differs from a permanent magnet because
- **a)** it does not have domains
- **b)** its magnetism lasts only a short time
- **c)** it is made of soft steel    **d)** it never has alloys

a   2. Most permanent magnets
- **a)** are made of steel    **b)** can never break
- **c)** are made of soft iron    **d)** lose magnetism easily

c   3. Steel, alnico, and Permalloy
- **a)** all become magnets easily    **b)** only make temporary magnets
- **c)** are alloys that can make magnets    **d)** lose their magnetism easily

a   4. Soft iron
- **a)** makes temporary magnets    **b)** makes permanent magnets
- **c)** is a temporary alloy    **d)** is always magnetic

d   5. In a temporary magnet, the domains
- **a)** cannot act as tiny magnets    **b)** cancel one another out
- **c)** never line up in the same direction    **d)** can change the way they line up

## WORD SCRAMBLE

*Below are several scrambled words you have used in this Lesson. Unscramble the words and write your answers in the spaces provided.*

1. MANTENREP                        permanent

2. MERATYPOR                     temporary

3. LEMPRYLOA                      Permalloy

4. LESET                                 steel

5. SROGNT                              strong

## REACHING OUT

Both soft iron and steel contain iron. What other metals does steel contain? Do some research to find out.

_____

_____

_____

_____

_____

# How can you make a magnet by induction?

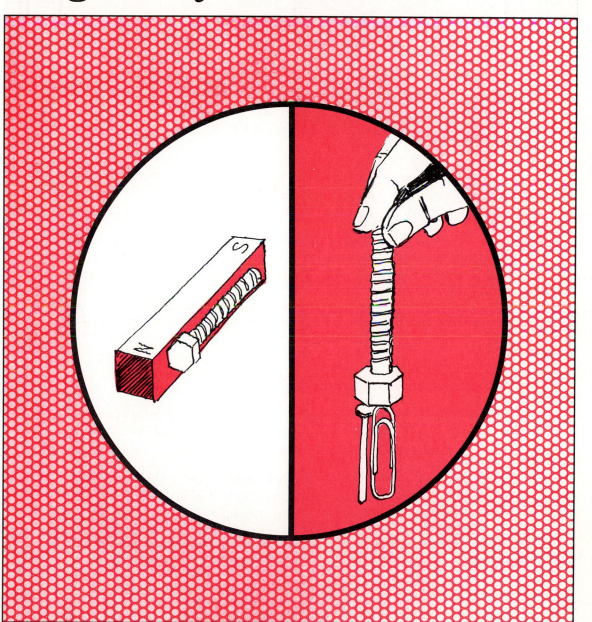

**magnetic induction:** the use of a magnetic field to make a magnetic substance into a magnet

# LESSON 29 | How can you make a magnet by induction?

You don't have to touch a fire to get burned. You can get burned just by being near a fire—heat energy spreads out. It can be felt at a distance from the flame.

Magnetic energy also spreads out. The energy can work at a <u>distance</u> from a magnet. Because of this, a magnetic substance can become a magnet just by being <u>near</u> a magnet.

This way of making a magnet is called **induction** [in DUCK shun]. Magnetic induction works best in a strong magnetic field. Induction works less well in a weak magnetic field. There can be no induction outside the magnetic field.

Both temporary and permanent magnets can be made by induction. The kind you get depends on two things:

1. the kind of metal
2. the strength of the magnetic field.

Iron can become a temporary magnet only. Iron can be magnetized by induction very easily. Even a weak magnetic field will work. But iron does not keep its magnetism. It loses it when the magnetic field is removed.

Steel can become a permanent magnet only. Steel cannot be easily magnetized by induction. A very strong magnetic field is needed. But steel keeps its magnetism once it does become a magnet.

The strong magnetic fields needed to make permanent magnets come from electricity.

This diagram shows a magnet, its magnetic field, and four pieces of iron. *Study the diagram. Then answer the questions.*

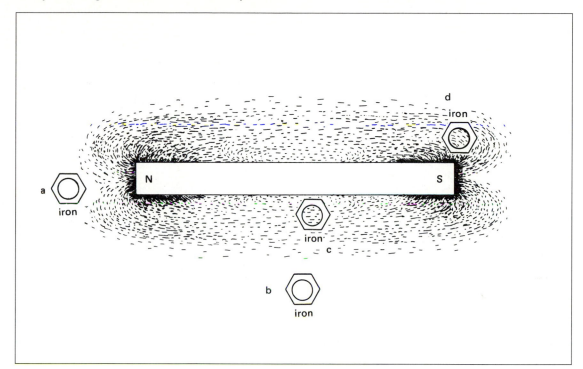

**Figure A**

1. **a)** Which iron pieces are inside the magnetic field? _____c, d_____

   **b)** Which iron pieces are outside the magnetic field? _____a, b_____

2. **a)** Which iron pieces have become magnets? _____c, d_____

   **b)** Which iron pieces have not become magnets? _____a, b_____

3. Look at the pieces that have become magnets.

   **a)** Which one is stronger? _____d_____

   **b)** Which one is weaker? _____c_____

4. The iron pieces in question #3 have become magnets by

   _____induction_____ .
   contact, induction, stroking

5. Magnetic induction takes place _____inside_____ a magnetic field.
   inside, outside

# STUDYING MAGNETIC INDUCTION

## What You Need (Materials)

magnet
large iron nail
small steel tacks

## How To Do the Experiment (Procedure)

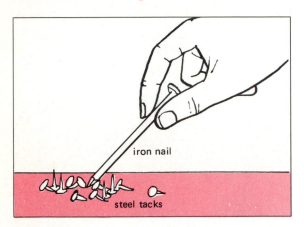

**Figure B**

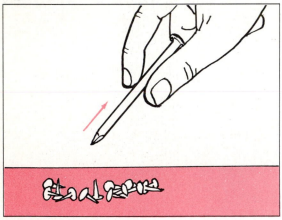

**Figure C**

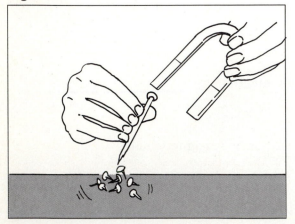

**Figure D**

Hold the end of the nail in the pile of tacks (Figure B).

Lift the nail (Figure C).

1. The nail __does not__ lift any
   tacks.    <sub>does, does not</sub>

2. This shows that the nail

   __is not__ a magnet.
   <sub>is, is not</sub>

Place the nail in the tacks again.
With your other hand, hold the magnet close to the head of the nail. (Careful, don't let them touch.)

Lift the nail and magnet together (Figure D). (Careful, keep the distance between the nail and the magnet.)

3. The nail __does__ lift
   tacks.    <sub>does, does not</sub>

4. The nail __has__ become
   a magnet.    <sub>has, has not</sub>

5. The magnet and nail

   __are not__ touching.
   <sub>are, are not</sub>

6. The nail __is__ in the
      <sub>is, is not</sub>

   magnetic field of the magnet.

7. By what method has the nail become

   a magnet? __induction__

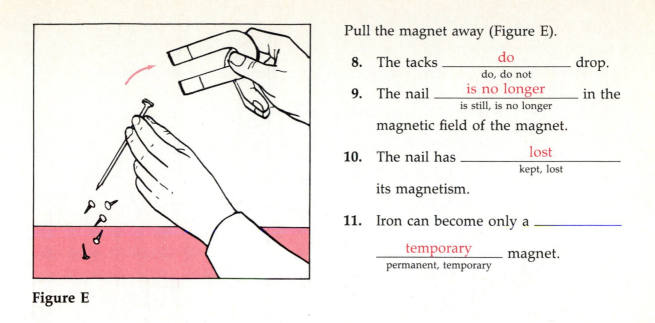

Pull the magnet away (Figure E).

8. The tacks _____do_____ drop.
   do, do not

9. The nail ___is no longer___ in the
   is still, is no longer
   magnetic field of the magnet.

10. The nail has _____lost_____
    kept, lost
    its magnetism.

11. Iron can become only a _____

    ___temporary___ magnet.
    permanent, temporary

**Figure E**

## FILL IN THE BLANK

*Complete each statement using a term or terms from the list below. Write your answers in the spaces provided. Some words may be used more than once.*

induction          contact          temporary
does not           weak             electricity
very strong        permanent        stroking
very easily

1. Two ways to make a magnet by touching are by ___contact___ and by
   ___stroking___ .

2. Magnetism at a distance is called ___induction___ .

3. Both ___temporary___ and ___permanent___ magnets are made by induction.

4. Iron can become only a ___temporary___ magnet.

5. Iron becomes a magnet ___very easily___ .

6. Even a ___weak___ magnetic field can make iron a magnet by induction.

7. Induction makes steel into a ___permanent___ magnet.

8. Steel ___does not___ become a magnet by induction easily.

9. A ___very strong___ magnetic field is needed to make steel a permanent magnet by induction.

10. Very strong magnetic fields come from ___electricity___ .

## TRUE OR FALSE

*In the space provided, write "true" if the sentence is true. Write "false" if the sentence is false.*

False     **1.** "Stroking" is magnetism at a distance.

False     **2.** "Contact" magnetism is magnetism at a distance.

True     **3.** "Induction" is magnetism at a distance.

True     **4.** Induction works only inside a magnetic field.

True     **5.** Iron can become a magnet by induction.

False     **6.** Iron becomes a permanent magnet.

True     **7.** Iron becomes a magnet easily.

True     **8.** Steel can become a magnet by induction.

False     **9.** Steel becomes a magnet easily by induction.

False     **10.** Steel becomes a temporary magnet.

## WORD SEARCH

*The list on the left contains words that you have used in this Lesson. Find and circle each word where it appears in the box. The spellings may go in any direction: up, down, left, right, or diagonally.*

MAGNET
IRON
ALLOY
ALNICO
NICKEL
COBALT
LODESTONE
DOMAIN

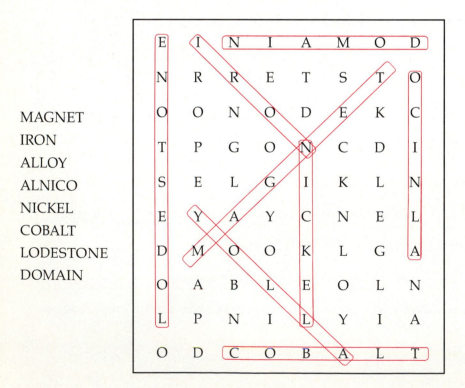

# What is an electromagnet?

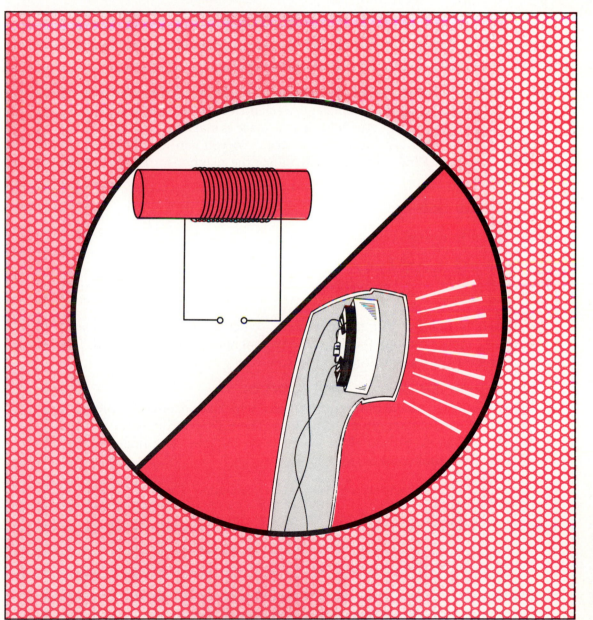

**electromagnet:** a temporary magnet made by using electrical current

# LESSON 30 | What is an electromagnet?

You may have seen a machine in a junk yard that lifts whole cars. This is a huge electromagnet. Did you know that there is a tiny electromagnet inside your telephone? It helps you hear people speak.

Just what is an **electromagnet?** It is a temporary magnet that gets its magnetism from electricity. Long ago, a scientist discovered that an electric current gives off a magnetic field. If you place a magnetic substance in a magnetic field, you can induce magnetism. So you can use electricity to make a magnet. When the electricity stops, the magnetism in an electromagnet stops.

Three things are needed to make an electromagnet:

1. a soft iron core

2. a coil of insulated wire

3. a source of electricity

A switch is helpful, but not necessary. When you connect the parts, electricity moves through the wire. The current makes a magnetic field. The field magnetizes the soft iron core by induction.

## THE STRENGTH OF ELECTROMAGNETS

Electromagnets come in different strengths, which are needed for different jobs. Weak electromagnets do small jobs. The tiny electromagnet in your telephone is weak. Strong electromagnets do big jobs. The huge electromagnet in the junk yard is strong.

How can you change the strength of an electromagnet? There are two ways:

1. change the number of coils the wire makes around the soft iron core

2. change the strength of the electric current

For example, to make an electromagnet stronger, you could wind more wire around the core, or increase the amount of electric current. There is a limit to how strong you can make an electromagnet. Once the core has the most magnetism it can take, it cannot be made stronger.

# WHAT DO THE PICTURES SHOW?

*Look at each picture. Then answer the questions.*

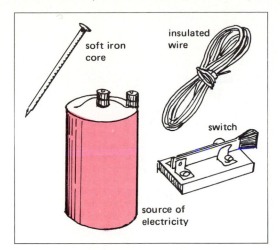

**Figure A**

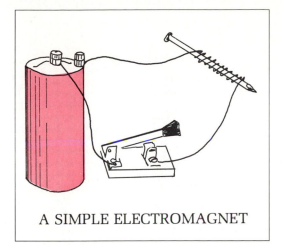

A SIMPLE ELECTROMAGNET

**Figure B**  *A simple electromagnet*

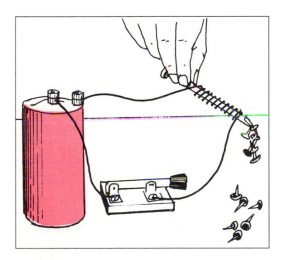

**Figure C**

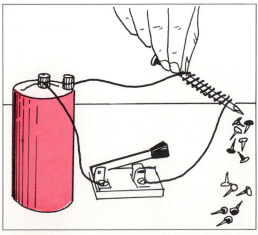

**Figure D**

1. Which of the parts in Figure A must an electromagnet have?

   <u>soft iron core, insulated wire, source</u>

   <u>of electricity</u>

2. Which part is helpful, but not necessary? <u>switch</u>

*Look at Figure C.*

3. This circuit is <u>complete</u>.
   complete, incomplete

4. Electricity <u>is</u>
   is, is not
   moving through the wire.

5. The iron core <u>has</u>
   has, has not
   become a magnet.

*Look at Figure D.*

6. If you open the switch, the tacks
   <u>drop</u>.
   drop, do not drop

7. An electromagnet is a
   <u>temporary</u> magnet.
   temporary, permanent

A teacher in Denmark discovered by accident that electricity makes a magnetic field. In 1819, Hans Oersted put a compass near an electric current.

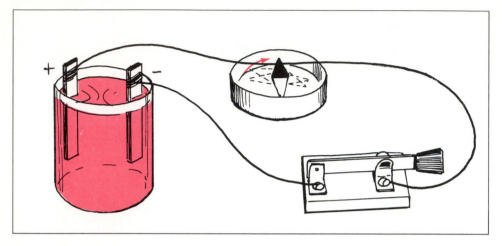

**Figure E**

When the current was on, the compass needle moved. It turned toward the wire.

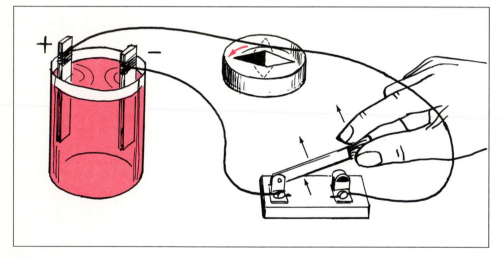

**Figure F**

When the current was off, the compass needle moved back to where it had been. Oersted tried this again and again. The same thing happened each time.

1. What made the compass needle turn? <u>The magnetic field made by the current in the wire.</u>

2. What did Oersted prove? <u>He proved that an electric current gives off a magnetic field.</u>

## FILL IN THE BLANK

*Complete each statement using a term or terms from the list below. Write your answers in the spaces provided. Some words may be used more than once.*

| | | |
|---|---|---|
| number of coils | strength of the current | coil of insulated wire |
| soft iron core | a magnetic field | electromagnet |
| induction | | source of electricity |

1. An _____electromagnet_____ is a temporary magnet.

2. An electromagnet becomes a magnet by _____induction_____ .

3. To make an electromagnet, you need: a _____soft iron core_____ , a _____coil of insulated wire_____ ,

   and a _____source of electricity_____ .

4. A current of electricity gives off _____a magnetic field_____ .

5. You can change the strength of an electromagnet by changing _____the number of coils_____

   _____the strength of the current_____ .

## TRUE OR FALSE

*In the space provided, write "true" if the sentence is true. Write "false" if the sentence is false.*

_____False_____ **1.** An electromagnet is a permanent magnet.

_____True_____ **2.** Soft iron loses its magnetism easily.

_____True_____ **3.** The core of an electromagnet is soft iron.

_____True_____ **4.** When an electromagnet is connected, the core has a north and south pole.

_____False_____ **5.** A magnetic field surrounds every wire.

## MATCHING

*Match each term in Column A with its description in Column B. Write the correct letter in the space provided.*

| Column A | | Column B |
|---|---|---|
| _____b_____ **1.** | soft iron core, coil of insulated wire, source of electricity | **a)** turns towards magnetic field |
| | | **b)** parts of an electromagnet |
| _____e_____ **2.** | magnetic field | **c)** discovered that electricity gives off magnetism |
| _____d_____ **3.** | strength of an electromagnet | |
| | | **d)** depends on current strength and number of coils |
| _____a_____ **4.** | compass | |
| _____c_____ **5.** | Oersted | **e)** given off by electricity |

# EXPERIMENTING WITH ELECTROMAGNETS

Try this. You will see how an electromagnet can be made stronger.

**What You Need (Materials)**

two 1½-volt dry cells
two large iron nails
insulated wire
switch
small steel tacks

**How To Do the Experiment (Procedure)**

1. Hook up your electromagnet. The chart below shows four different ways.

2. Test each hook up.

**Figure G**   *One dry cell*

**Figure H**   *Two dry cells*

## Observations

*Count how many tacks the electromagnet lifts. Write the numbers on the chart.*

| Number of Wire Turns | Number of Dry Cells | Number of Tacks Picked Up |
|---|---|---|
| 20 | 1 | |
| 40 | 1 | |
| 20 | 2 | |
| 40 | 2 | |

Design and test your own electromagnet. *Write what you have learned in the space below.*

answers will vary

increasing the number of coils or the current strength will make the magnet stronger

194

# What is a transformer?

31

transformer: a device that changes the voltage of alternating current
direct current: electric current flowing in one direction
alternating current: electric current that reverses direction of flow

# LESSON 31 | What is a transformer?

Some electric passenger trains need huge electromotive force to turn their wheels—as much as 11,000 volts. An electric toy train uses only about 18 volts. The wires in most homes carry about 115 volts. Some appliances need high voltage to work. Some need low voltage.

Sometimes voltage must be made stronger; sometimes it must be made weaker to suit different devices.

How can we change voltage? By using a **transformer.** There are large and small transformers. But they all work the same way.

A transformer has three main parts: (Figure A).

- a soft iron core, and

- two different coils of insulated wire wrapped around the core.

One of these coils is called the <u>primary coil</u>. The primary coil is connected to the electricity coming in.

The other coil is called the <u>secondary coil</u>. The secondary coil is connected to the appliance.

There are two kinds of transformers—<u>step-up</u>, and <u>step-down</u>.

**I.** A step-up transformer increases voltage.

In a step-up transformer, the secondary coil is wrapped around the core more times than the primary coil.

**II.** A step-down transformer decreases voltage.

In a step-down transformer, the primary coil is wrapped around the core more times than the secondary coil.

There are two kinds of electric current—**direct current** (DC) and **alternating current** (AC). Transformers work only with alternating current.

## UNDERSTANDING TRANSFORMERS

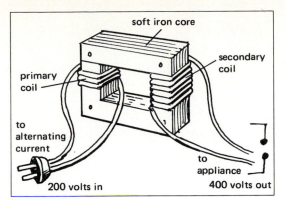

**Figure A**  *A step-up transformer*

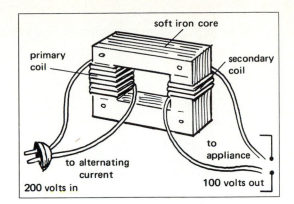

**Figure B**  *A step-down transformer*

This step-up transformer has 3 turns in the primary coil.

The secondary coil has 6 turns. This is twice the turns as the primary coil. It <u>doubles</u> the voltage.

For example: If you start out with 200 volts, you end up with 400 volts.

This step-down transformer has 6 turns in the primary coil.

The secondary coil has 3 turns. This is one half the turns as the primary coil. It makes the voltage half as strong.

For example: If you start out with 200 volts, you end up with 100 volts.

## DIRECT CURRENT AND ALTERNATING CURRENT

### In DIRECT CURRENT:

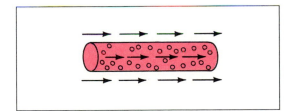

**Figure C**

**a)** The electricity moves through the circuit in <u>one direction only</u>.

**b)** It does <u>not</u> move back and forth.

**c)** Direct current does not stop and go, stop and go. REMEMBER THIS.

### In ALTERNATING CURRENT:

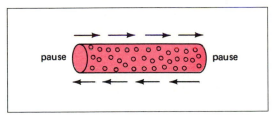

**Figure D**

**a)** The electricity moves back and forth many times a second.

**b)** Each time the electricity changes direction, it <u>stops for a moment</u>. It happens very fast.

**c)** Remember—alternating current stops and goes, stops and goes, stops and goes.

## A TRANSFORMER CHANGES ONLY ALTERNATING CURRENT

Most people all over the world use alternating current.

*Figures E through H show transformers. Study each one. Then answer the questions next to the figure.*

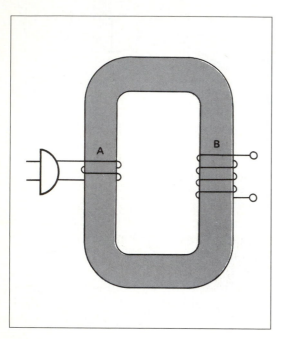

**Figure E**

1. In Figure E, the primary coil is
   <u>     A     </u>.
   A, B

2. The secondary coil is <u>   B   </u>.
   A, B

3. How many turns does the primary
   coil have? <u>   2   </u>

4. How many turns does the secondary
   coil have? <u>   4   </u>

5. The voltage is being made
   <u>   stronger   </u>.
   stronger, weaker

6. This is a <u>  step-up  </u>
   step-up, step-down
   transformer.

7. If "A" has 100 volts, how many volts
   does "B" have? <u>200</u>

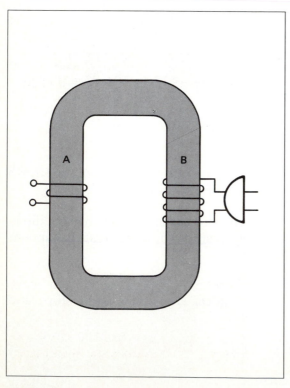

**Figure F**

1. In Figure F, the primary coil is
   <u>     B     </u>.
   A, B

2. The secondary coil is <u>   A   </u>.
   A, B

3. How many turns does the primary
   coil have? <u>   4   </u>

4. How many turns does the secondary
   coil have? <u>   2   </u>

5. This is a <u>  step down  </u>
   step-up, step-down
   transformer.

6. The voltage is being made
   <u>   weaker   </u>.
   stronger, weaker

7. If "B" has 100 volts, how many volts
   does "A" have? <u>50</u>

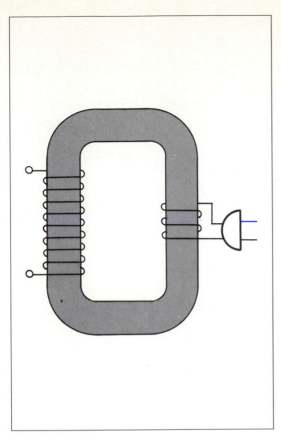

**Figure G**

1. In Figure G, how many turns are there

   a) in the primary coil? _____3_____

   b) in the secondary coil? _____9_____

2. The primary coil is connected to the

   _____electric current_____ .
   electric current, appliance

3. The secondary coil is connected to

   the _____appliance_____ .
   electric current, appliance

4. This is a _____step-up_____
   step-up, step-down

   transformer.

5. The voltage is being made

   _____stronger_____ .
   stronger, weaker

6. If the primary coil has 100 volts, how many volts does the secondary

   coil have? _____300_____

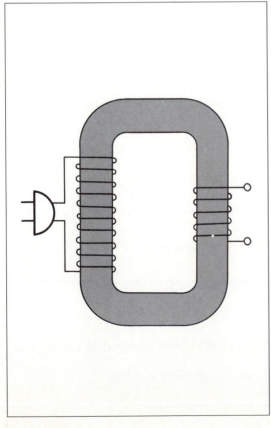

**Figure H**

1. In Figure H, how many turns are there

   a) in the primary coil? _____10_____

   b) in the secondary coil? _____5_____

2. The primary coil is connected to the

   _____electric current_____ .
   electric current, appliance

3. The secondary coil is connected to

   the _____appliance_____ .
   electric current, appliance

4. This is a _____step-down_____
   step-up, step-down

   transformer.

5. The voltage is being made

   _____weaker_____ .
   stronger, weaker

6. If the primary coil has 100 volts, how many volts does the secondary

   coil have? _____50_____ .

# FILL IN THE BLANK

*Complete each statement using a term or terms from the list below. Write your answers in the spaces provided.*

two coils of insulated wire        step-up                without stopping
stops and goes                     direct                 weaker
step-down                          soft iron core         transformer
alternating                        stronger

1. There are two kinds of electric currents. They are _____direct_____ current, and _____alternating_____ current.

2. Direct current moves _____without stopping_____ .

3. Alternating current _____stops and goes_____ many times every second.

4. A _____transformer_____ changes voltage.

5. A transformer works only with _____alternating_____ current.

6. The important parts of a transformer are: a _____soft iron core_____ and _____2 coils of insulated wire_____ .

7. A step-up transformer makes voltage _____stronger_____ .

8. A step-down transformer makes voltage _____weaker_____ .

9. A _____step-up_____ transformer has more turns in the secondary coil than in the primary coil.

10. A _____step down_____ transformer has more turns in the primary coil than in the secondary coil.

# MATCHING

*Match each term in Column A with its description in Column B. Write the correct letter in the space provided.*

**Column A**

___b___ 1. step-up transformer

___d___ 2. step-down transformer

___a___ 3. soft iron core and two sets of coils

___e___ 4. soft iron

___c___ 5. alternating current

**Column B**

a) important parts of a transformer

b) makes voltage stronger

c) stops and goes many times a second

d) makes voltage weaker

e) transformer core

## TRUE OR FALSE

*In the space provided, write "true" if the sentence is true. Write "false" if the sentence is false.*

False     **1.** A transformer changes amperes.

False     **2.** The core of a transformer is made of steel.

False     **3.** The primary coil of a transformer is connected to the appliance.

True     **4.** The secondary coil of a transformer is connected to the appliance.

False     **5.** A transformer only raises voltage.

True     **6.** A step-up transformer makes voltage stronger.

True     **7.** A step-down transformer makes voltage weaker.

False     **8.** Direct current starts and stops many times a second.

True     **9.** Alternating current starts and stops many times a second.

True     **10.** A transformer works only with alternating current.

## COMPLETE THE CHART

*Complete the chart by filling in the missing information. Line 1 has been filled in for you.*

|  | Step-up Transformer | Step-down Transformer | Primary Coil Turns | Secondary Coil Turns | Primary Coil Voltage | Secondary Coil Voltage |
|---|---|---|---|---|---|---|
| **1.** | ✔ |  | 20 | 40 | 10 | 20 |
| **2.** | ✔ |  | 30 | 60 | 20 | 40 |
| **3.** |  | ✔ | 25 | 5 | 50 | 10 |
| **4.** | ✔ |  | 1 | 10 | 5 | 50 |
| **5.** |  | ✔ | 10 | 2 | 50 | 10 |
| **6.** | ✔ |  | 20 | 60 | 100 | 300 |
| **7.** | ✔ |  | 6 | 24 | 50 | 200 |
| **8.** | ✔ |  | 4 | 20 | 5 | 25 |
| **9.** |  | ✔ | 100 | 10 | 50 | 5 |
| **10.** |  | ✔ | 500 | 50 | 300 | 30 |

## REACHING OUT

A powerhouse generator may produce more than 22,000 volts. Huge step-up transformers boost it to nearly 350,000 volts.

1. Why is the voltage raised so much? _Because powerhouses supply many homes and businesses with electric power, which must be sent over long distances. Higher voltage allows the current to travel farther and to more places to be used._

2. What happens to the voltage before it reaches your home? _Step-down transformers reduce voltage before it goes to the power lines connected to customers' homes and businesses._

# What is an induction coil?

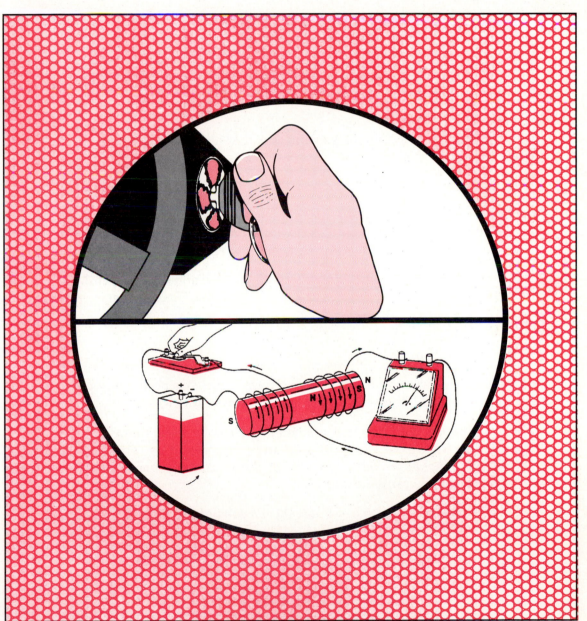

**induction coil:** a device that increases the voltage of direct current

# LESSON 32 | What is an induction coil?

Cars, trucks, and buses run on gasoline. But they also need electricity. Each one needs about 20,000 volts!

Most car batteries give only 12 volts. This is not nearly strong enough.

Why not use a transformer to raise the voltage? This sounds like a good idea, but it won't work. A transformer works only with <u>alternating</u> current. And a car battery gives only <u>direct</u> current. Something else is needed.

What can boost the voltage of direct current? We use an **induction coil.** An induction coil and a transformer are very much alike. Each one has

• a soft iron core, and

• two coils of insulated wire.

But there is one important difference. A transformer has no moving parts. An induction coil has an extra part that is always moving.

The extra part is a <u>switch</u> that opens and closes by itself many times a second. This makes the electricity stop-and-go, stop-and-go. The on-and-off switching makes the direct current act like alternating current. Because of this, the voltage can be changed.

The induction coil in a car can boost the 12 volts of the battery to 20,000 volts.

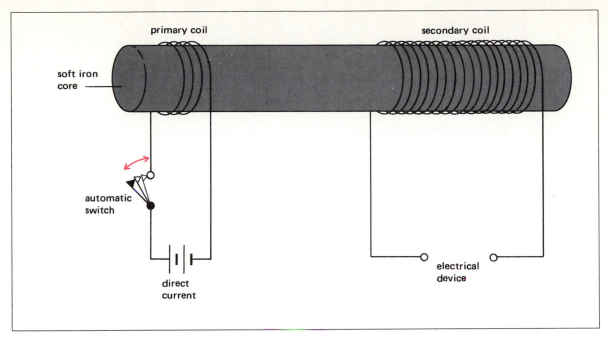

**Figure A** *An induction coil*

# FILL IN THE BLANK

*Complete each statement using a term or terms from the list below. Write your answers in the spaces provided. Some words may be used more than once.*

| | | |
|---|---|---|
| direct current | alternating current | automatic switch |
| induction coil | direct | transformer |
| alternating | | |

1. There are two kinds of electricity; _____direct_____ current, and

    _____alternating_____ current.

2. DC stands for _____direct current_____.

3. AC stands for _____alternating current_____.

4. Electricity in schools and homes is usually _____alternating_____ current.

5. _____Direct_____ current moves without stopping.

6. _____Alternating_____ current starts and stops many times every second.

7. A _____transformer_____ changes the voltage of alternating current.

8. An _____induction coil_____ changes the voltage of direct current.

9. Batteries give only _____direct_____ current.

10. An induction coil has a part that a transformer does not have. That part is an

    _____automatic switch_____.

## MATCHING

Match each term in Column A with its description in Column B. Write the correct letter in the space provided.

|  | Column A |  | Column B |
|---|---|---|---|
| b | 1. induction coil | a) | toy electric trains |
| e | 2. car battery | b) | change voltage of DC |
| a | 3. transformer | c) | circuit in the home |
| c | 4. AC | d) | induction coil's moving part |
| d | 5. switch | e) | source of DC |

## WORD SEARCH

The list on the left contains words that you have used in this Lesson. Find and circle each word where it appears in the box. The spellings may go in any direction: up, down, left, right, or diagonally.

NORTH
SOUTH
INDUCTION
CORE
CURRENT
BATTERY
WIRE
COIL

```
H  S  O  E  B  C  U  T  N  H
P  D  E  C  A  H  T  U  O  S
W  N  O  I  T  C  U  D  N  I
R  I  W  E  T  O  T  N  O  W
L  N  R  I  E  R  S  U  R  H
W  C  U  R  R  E  N  T  T  O
C  I  E  W  Y  E  I  E  H  Y
```

## REACHING OUT

An induction coil can make a flashlight burn brighter. Why is it not used?

The use of an induction coil requires a way to switch the power on and off. An automatic switch needs an additional source of energy.

206

# How does an electrical generator work?

**generator:** a machine that changes one form of energy into electrical energy
**galvanometer:** a device that measures weak electric current

# LESSON 33 | How does an electrical generator work?

Magnetism and electricity are not the same, but they are related. You learned that electricity can make magnetism. The opposite is also true: magnetism can produce electricity.

With a wire and a magnet, you can make electricity. Moving a magnet back and forth inside a coil of wire will make an electric current. Moving the coil of wire back and forth around the magnet will also make an electric current. This is what happens in an electrical **generator.**

### INSIDE A GENERATOR

A generator is a machine that changes one form of energy into electrical energy. When a coil of wire turns between the poles of a permanent magnet, the wire moves through a magnetic field. When any conductor cuts across lines of magnetic force, electrons move in the conductor. The coil of wire is a conductor. Electric current is the movement of electrons. So as the wire turns in the magnetic field, electric current flows through it.

The faster the coil turns in the magnetic field, the stronger the electric current that is generated. The stronger the magnet in a generator, the stronger the electric current that is made.

Something else has to supply energy to turn the coil of wire in the magnetic field. That is why we say that a generator changes one form of energy into electrical energy.

## INSIDE A GENERATOR

Name the essential parts of a generator:

a coil of insulated wire    a permanent magnet

*Look at Figure A.*

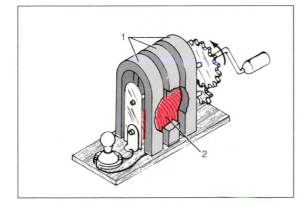

**Figure A**

What are the parts labeled 1? __permanent magnets__

What is the part labeled 2? __a coil of wire__

Must a generator have only one permanent magnet? __No__

How many magnets does the generator in Figure A have? __3__

You can make a simple generator with a coil of insulated wire, a bar magnet, and a **galvanometer.** A galvanometer is a device that measures weak electric current.

*Look at Figure B.*

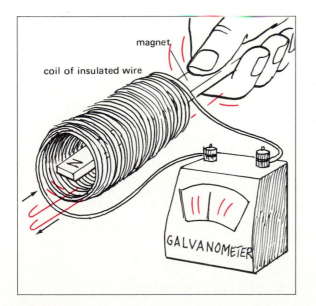

**Figure B**

1. In this simple generator, the

   ____magnet____ is moving.
   <sub>magnet, coil</sub>

2. If the coil would move instead, the

   electricity would __be the same__.
   <sub>stop, be the same</sub>

3. If you move the coil or magnet

   slower, you get ____weaker____
   electricity.   <sub>stronger, weaker</sub>

4. If you move the coil or magnet

   faster, you get ____stronger____
   electricity.   <sub>stronger, weaker</sub>

5. If you use a stronger magnet, you get

   ____stronger____ electricity.
   <sub>stronger, weaker</sub>

## FILL IN THE BLANK

Complete each statement using a term or terms from the list below. Write your answers in the spaces provided. Some words may be used more than once.

move                          magnetism                  permanent magnet
electricity                   weak                       galvanometer
turn the wire coil faster     use a stronger magnet      generator
coil of insulated wire

1. Electricity can make _____ magnetism _____ .

2. Magnetism can be used to make _____ electricity _____ .

3. The machine that makes electricity is called a _____ generator _____ .

4. The necessary parts of an electric generator are a _____ coil of insulated wire _____ and

   a _____ permanent magnet _____ .

5. To make electricity, either the wire or the magnet must _____ move _____ .

6. In most generators, the _____ coil of wire _____ moves.

7. A _____ galvanometer _____ measures weak electricity.

8. Weak magnets can make only _____ weak _____ electricity.

9. Two ways to make stronger electricity are _____ turn the wire coil faster _____

   and _____ use a stronger magnet _____ .


## MATCHING

Match each term in Column A with its description in Column B. Write the correct letter in the space provided.

| Column A | Column B |
|---|---|
| __b__  1. generator | a) cuts across lines of magnetic force |
| __d__  2. magnet and coil of insulated wire | b) produces electrical energy |
| __c__  3. galvanometer | c) measures weak voltage |
| __e__  4. electricity | d) parts of a generator |
| __a__  5. moving coil of wire | e) moving electrons |

# What can be used to power a generator?

34

NUCLEAR
FISSION

Large
nucleus

Smaller nuclei

Energy

# LESSON 34 | What can be used to power a generator?

A generator needs energy to turn the wire coils in the magnetic field of the permanent magnets. What can supply this energy?

Let's examine three common energy sources.

## HYDROELECTRIC POWER

Hydroelectric power is generated by the force of moving water. In some places, like Niagara Falls, natural waterfalls provide the force needed to turn the generators. In other places, controlled release of water from a dam on a river supplies the energy.

*PROS*—Hydroelectric power is renewable. It is renewable because the rivers continue to flow without our help. It is also nonpolluting. Hydroelectric power can be less expensive than other kinds.

*CONS*—Hydroelectric power cannot be generated in all areas. The geography must be just right. A supply of moving water must always be available. In some locations, it's not possible to build a dam on a river.

## FOSSIL FUELS

Coal, oil or petroleum, and natural gas are fossil fuels. They are taken out of the ground and burned for their energy. Most generators are powered by fossil fuel.

*PROS*—Fossil fuels are in good supply. Furthermore, fossil fuels are well understood. They have been used for a long time.

*CONS*—Fossil fuels are not renewable and cause pollution. Pollutants from fossil fuels contribute to acid rain and the greenhouse effect, or global warming. Coal is the worst polluter when burned.

## NUCLEAR ENERGY

Nuclear power comes from energy released by splitting atoms. Splitting atoms is called nuclear fission.

*PROS*—A very small amount of nuclear fuel can produce a tremendous amount of energy. Nuclear energy does not emit pollution that causes acid rain or global warming. There is a plentiful supply of nuclear fuel.

*CONS*—Leftover materials from nuclear fuel, called radioactive wastes, are dangerous enough to cause illness or death. And there does exist the danger of a nuclear accident that can release radioactive waste into the environment.

# ELECTRIC POWER FROM DIFFERENT ENERGY SOURCES

Electrical generators in the U.S. are powered by several energy sources. The pie graph below in Figure A shows the approximate percent share of each source. Next to each pie section, write the power source it stands for—and the *percent*.

*Choose from the following:*

Coal 56%                     Natural gas 9%
Nuclear 19.0%                Petroleum (oil) 6%
Hydroelectric (water) 9%     Others less than 1%

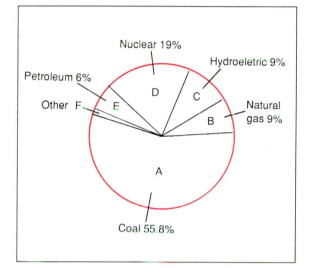

**Figure A**

*Now answer these questions.*

1. Which power source generates *most* electricity? _____coal_____

2. Which two power sources generate the least? __Petroleum and other_____

3. Which are the fossil fuels? __Petroleum, coal, and natural gas_____

4. Which fossil fuel produces

   **a)** the most electricity? _____coal_____

   **b)** the least electricity? _____petroleum_____

5. Altogether, fossil fuels supply the energy for _____71_____ percent of our nation's electrical needs.

6. This is closest to what fraction?

   **a)** ¼    **b)** ⅓    **c)** ½    **d)** ¾ _____¾_____

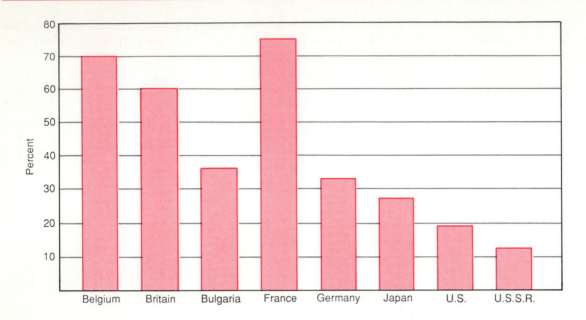

**Figure B**

Figure B above shows eight countries. It also indicates the approximate percentage of their electrical needs supplied by nuclear power.

*Study the figure. Answer these questions. Next to each country, write the percentage of its electrical needs supplied by nuclear energy. Write percent as a symbol (%).*

Belgium __60%__  Germany __33%__  France __75%__

Britain __20%__  Japan __27%__  U.S.S.R. __12%__

Bulgaria __35%__  U.S. __19%__

Which country obtains the *greatest* percent of its electrical needs from nuclear power?

__France__

Which country obtains the smallest percent of its electrical needs from nuclear power?

__U.S.S.R.__

The United States has *more than twice* the number of nuclear reactors than France. Yet, France supplies 75% of its electrical needs with nuclear energy—while the U.S. supplies only 21%.

How can this be explained? __The United States uses much more electricity than France.__

__The U.S. also has a much larger population, and therefore needs more electricity than__

__France.__

## FILL IN THE BLANK

*Complete each statement using a term or terms from the list below. Write your answers in the spaces provided. Some words may be used more than once.*

|  |  |  |
|---|---|---|
| fossil fuels | generators | petroleum |
| natural gas | location | nuclear |
| coal | water |  |

1. Large and continuous amounts of electricity can be supplied only by
   _____generators_____.

2. Most generators in the United States are powered by ____fossil fuels____.

3. The fossil fuels are _____natural gas_____ , _____coal_____ and
   _____petroleum_____.

4. The fossil fuel most used to power American generators is _____coal_____.

5. "Hydro" refers to _____water_____.

6. Hydroelectric generation is limited by _____location_____.

7. The fuel that uses small amounts to produce the most energy is _____nuclear_____
   fuel.

8. The most polluting fuel is _____coal_____.

9. The only major fuel that does not contribute to global warming or acid rain is
   _____nuclear_____ fuel.

## MATCHING

*Match each term in Column A with its description in Column B. Write the correct letter in the space provided.*

| Column A | Column B |
|---|---|
| ___d___ 1. coal | **a)** not renewable |
| ___e___ 2. hydroelectric power | **b)** danger from nuclear power |
| ___a___ 3. fossil fuels | **c)** comes from splitting atoms |
| ___c___ 4. nuclear power | **d)** contributes to acid rain |
| ___b___ 5. radioactive waste | **e)** nonpolluting |

## TRUE OR FALSE

*In the space provided, write "true" if the sentence is true. Write "false" if the sentence is false.*

   True       **1.** Large generators supply most of our electricity.

   False       **2.** Hydroelectric power can be generated anywhere.

   True       **3.** Generators powered by fossil fuels and nuclear energy can be built in many locations.

   True       **4.** Fossil fuels pollute.

   False       **5.** Nuclear power contributes to acid rain.

   False       **6.** Fossil fuels are renewable.

   False       **7.** Hydroelectric power leads to global warming.

   False       **8.** Any large body of water can power generators.

   True       **9.** Water is a renewable power source.

   True       **10.** Coal is the most polluting fuel.

## REACHING OUT

Find out what source of energy powers the generators that make your electricity. Contact your electric company for information. Find out where they get the fuel for their generators.

# What are alternate sources of energy?

35

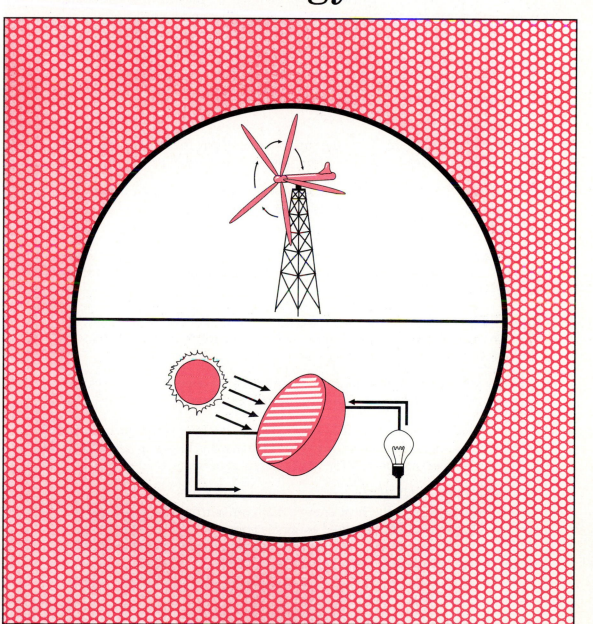

**geothermal:** relating to heat produced within the earth

# LESSON 35 | What are alternate sources of electricity?

The future is closer than you think. Before long, we may have to use <u>solar</u>, <u>wind</u>, and **geothermal** energy. They are <u>renewable</u> and almost <u>nonpolluting</u>. Today, these alternate sources of energy supply only a small fraction of our total need for electricity. But as time passes, new inventions and improved technology will make them more practical.

Let us briefly examine each one.

## SOLAR ENERGY
Solar refers to the energy from the sun. We do not pay for the sun. We must pay only for devices to capture, store, and put its energy to use.

Solar energy has a major drawback. The sun does not shine on any one place all the time. It is completely gone at night, and is largely blocked when it is cloudy.

Some parts of our country can benefit from solar energy more than others. Parts of the southwest can benefit most. The sun shines there more than 90% of daylight hours.

## WIND ENERGY
Wind is caused by the uneven heating of the earth's surface. For this reason, wind is considered indirect solar energy.

Wind is an ancient power source. It is now being used in new ways. As early as 1910 wind was used to turn electric generators.

The drawback of wind is that it is not steady. It blows stronger in some places than others. In fact, in some areas it hardly blows at all. Windy regions, however, are perfect for wind-driven generators.

## GEOTHERMAL ENERGY
Geothermal energy is the heat energy that comes from below the earth's surface. The amount of this stored energy is enormous. Geothermal energy comes from molten rock, and from materials in the earth's crust heated by radioactivity. Certain regions like Iceland, Mexico, New Zealand, and Japan are near volcanic areas. Geothermal energy has been used there for many years, mostly for heating.

Geothermal electric production is growing rapidly in several countries. In America, states such as Nevada, California, Hawaii, and Utah are the leaders in geothermal electric production. By the year 2000, California is expected to produce 25% of its electricity from geothermal energy.

# MORE ABOUT SOLAR ENERGY

There are three main types of solar energy systems—<u>passive</u>, <u>active</u> and <u>photovoltaics</u>. Two are used mainly for heating. One is used to make electricity.

## A. Passive Solar Energy System

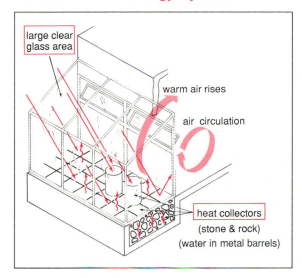

large clear glass area

warm air rises

air circulation

heat collectors
(stone & rock)
(water in metal barrels)

**Figure A**

A passive system is the simplest and least costly solar energy system. It is designed to capture, store, and then release the sun's energy. It is used mostly to heat homes and buildings. See Figure A.

1. A passive solar system has just two essential parts. Look at Figure A. Name these two parts. <u>large glass enclosed space</u> <u>heat collectors</u>

   Use your head in answering questions 2–3. Don't rush. Think them out carefully.

2. What happens during the daytime? <u>Warm air moves into the house from the glass enclosed area, as the sun heats up the heat collectors.</u>

3. What happens at night? <u>The heat collectors cool, and they warm the air around them as they do.</u>

## B. Active Solar Energy System

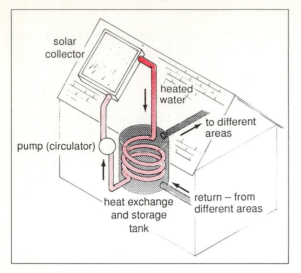

**Figure B**

Figure B shows an active solar system.

Active solar systems heat homes and buildings. They also heat water for washing, bathing and commercial use.

Active solar systems can also be used for cooling. But, as yet, they are not often used for this purpose.

4. Most active solar systems have one key part that no passive solar system has. Name

   that part. _____ a pump or circulator _____

5. Solar collectors of active solar systems are usually mounted on the ____roof____.

6. What substance is being heated by the solar collector? ____water____

## C. Photovoltaics (PV)

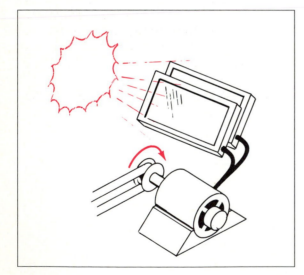

**Figure C**

Photovoltaics (PV) are also called solar cells. Solar cells change solar energy directly to electricity. Since 1958, instruments of almost every U.S. satellite have been powered by PV cells.

7. Most PV systems require many PV cells hooked up together. Why is this necessary?
   Each one alone does not provide enough power

8. Some students have a small device that is powered by a *single* PV cell. Chances are

   *you* own one. Name that device? _____ pocket calculator _____

# MORE ABOUT WIND ENERGY

**Figure D**

Wind can be trapped to do many useful jobs. For centuries wind filled the sails of ships and made them move. Windmills were used to grind grain, and to pump water from the ground. In the United States, windmills played an important role in the opening of the west. In the early 1900's, 6 million windmills were working in our nation. Now, in a few places such as California, windmills are working to power electrical generators.

1. Put on your thinking cap. What are the three main advantages of wind energy? <u>Low cost, Renewable, Non-polluting</u>

2. What is the main disadvantage of wind energy? <u>not a steady power source</u>

# MORE ABOUT GEOTHERMAL ENERGY

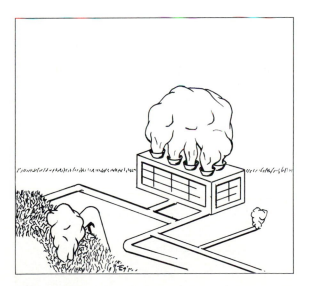

**Figure E**

Tapping into geothermal energy is not very difficult. Geologists determine the location of a geothermal "pocket." This is where heat rises from deep inside the earth toward the surface. A hole is drilled until it reaches the pocket. Pipes are inserted. Valves and pumps are connected.

Some geothermal pockets release *hot water*. Others release *steam*.

1. What might the hot water be used for? <u>heating or washing</u>

2. What might the steam be used for? <u>source of power for a generator</u>

3. Nature is constantly renewing geothermal energy. However, in some places, people may use this energy faster than nature can replace it.

   What, then, will happen? <u>Some places will lose the ability to tap geothermal energy from the ground.</u>

## FILL IN THE BLANK

*Complete each statement using a term or terms from the list below. Write your answers in the spaces provided. Some words may be used more than once.*

| | | |
|---|---|---|
| polluting | wind | passive |
| volcanic | electricity | sunny |
| solar | nonrenewable | active |
| geothermal | southwest | |

1.  Substitutes for fossil fuels are needed. This is because fossil fuels are

    __nonrenewable__ and also highly __polluting__ .

2.  Three of the best "alternate" energy sources are __solar__ ,

    __wind__ , and __geothermal__ energy.

3.  Solar energy works best in __sunny__ regions.

4.  All regions of the United States can benefit from solar energy. The region that can

    benefit most is the __southwest__ .

5.  The simplest solar energy system is the __passive__ solar system.

6.  The solar energy system that uses a circulator is the __active__ solar
    system.

7.  A photovoltaic cell changes __solar__ energy directly to

    __electricity__ .

8.  The energy source caused by the uneven heating of the earth's surface is

    __wind__ .

9.  Heat energy below the earth's surface is called __geothermal__ energy.

10. Most geothermal energy is found in __volcanic__ regions.

## REACHING OUT

Could solar and wind energy be used where you live? Find out how many days per year have sunshine where you live. The best wind speeds for making electricity are between 10 and 25 miles per hour. Find out how many days per year have wind speeds in this range where you live. Use a weather almanac or records from the weather bureau. Your librarian can help you.

# How can people conserve energy?

36

# LESSON 36 | How can people conserve energy?

In the future, renewable energy sources will replace fossil fuels. Energy will cost less. And it will be clean. Pollution from burning fuel will be a smaller problem.

However, we are not living in the future. The reality of the present is this:

- Most of our energy comes from fossil fuels.

- Fossil fuels cause very serious pollution.

- The cost of fossil fuels is high. And, it will continue to rise.

- Every bit of fossil fuel we use is GONE! Nature will not replace it for hundreds of millions of years. Can you wait that long?

For these reasons, we must learn to conserve. Energy conservation will save money. Even more important, it will reduce pollution.

What can YOU do to conserve energy? Many things! As a young person, you make many energy decisions every day. In addition, before long, you will be an adult—and on your own. Then, you will make decisions your parents now make—like choosing electrical appliances. Wise decisions can save much energy—and money.

For example: The use of room air conditioners is growing constantly. They offer great comfort. But air conditioners are expensive to run. They use a lot of electricity. A federal law requires every air conditioner displayed in an appliance store to carry an Energy Guide label, along with its Energy Efficiency Rating, or EER.

EER is a measure of how well an air conditioner cools compared to the amount of electricity it uses. Most EER ratings fall between 7 and 12. The higher the EER, the more efficient the unit is. When you buy, look for an EER of 9.0 or higher. 9.0 and above is considered very efficient.

Only room air conditioners are rated by EER. Many other high energy appliances, such as refrigerators, dishwashers, washing machines, and clothes dryers are rated in a different way. They are rated by their estimated yearly cost to use. For example, an energy guide label on an appliance might say that the product costs about $100.00 per year to use at home.

# UNDERSTANDING THE EER ENERGY GUIDE

Below is a sample Energy Efficiency Rating label.

*Study this label and check back in the reading as needed to answer the questions.*

**Figure A**

1. What do the initials EER stand for?

   <u>Energy Efficiency Rating</u>

2. What does EER measure?

   <u>how well an air conditioner cools</u>

   <u>compared to its electrical use.</u>

3. What is the usual EER range of values?

   <u>from 7 to 12</u>

4. The higher the EER the

   <u>    more    </u> efficient the
   less, more

   appliance is.

5. The higher the EER, the <u>    less    </u> expensive the appliance is to use.
   more, less

6. Which EER values are considered very efficient? <u>9 through 12</u>

7. Is the EER shown in Figure A among the very efficient? <u>No, it is a little less.</u>

8. How much would the air conditioner shown in the figure cost to run for:

   **a)** 750 hours in an 8¢/kilowatt hour area? <u>$66.00</u>

   **b)** 3,000 hours in a 4¢/kilowatt hour area? <u>$131.00</u>

9. The government tested many similar air conditioners to arrive at its 8.7 EER rating.

   The <u>most</u> efficient model rated <u>    1.5    </u> EER points <u>    higher    </u>
   (number) higher, lower

   than the labeled air conditioner.

10. The <u>least</u> efficient model rated <u>    3.0    </u> EER points <u>    lower    </u>.
    (number) higher, lower

11. Look at the label. On average, the EER rating of this air conditioner was

    <u>    higher    </u> than the other models tested.
    higher, lower

12. What on the label tells us how to compare this appliance to other models tested?

    <u>The black bar below the rating number shows the comparison of this model to other</u>

    <u>models.</u>

# ANOTHER ENERGY GUIDE LABEL

Energy guide labels for other kinds of electrical appliances show an estimated yearly cost to use. This gives an indirect value for EER. Study this example:

• Brand X refrigerator costs $100.00 per year to run.

• Brand Y is a similar refrigerator. But it costs *$145.00* to run—in the same energy-cost area.

If rated by EER which brand would have a higher rating? _____X_____
X, Y

The **national** *average* **rate** used to estimate the model's yearly operating cost.

The **Dollar Figure** is an estimate of the model's yearly operating cost.

The **Scale** shows the *range* of operating costs of similar models.

The ▼ shows where this model's operating costs fall in comparison to all other similar models.

The approximate **Yearly Cost Chart**. Match your utility rate with those on the chart to estimate your yearly rate.

(Name of Corporation)
Refrigerator     Model(s) AH503, AH504, AH507
Capacity: 23 Cubic Feet   Type of Defrost: Full Automatic

**ENERGYGUIDE**

Estimates on the scale are based on national average electric rate of 6.75¢ per kilowatt hour.

Only models with 22.5 to 24.4 cubic feet are compared in the scale

Model with lowest energy cost $92

**$124**
THIS ▼ MODEL
Estimated yearly energy cost

Model with highest energy cost $179

Your cost will vary depending on your local energy rate and how you use the product. The energy cost is based on U.S. Government standard tests.

**How much will this model cost you to run yearly?**

| | | Yearly cost |
|---|---|---|
| | | Estimated yearly $ cost shown below |
| Cost per kilowatt hour | 2¢ | $ 36 |
| | 4¢ | $ 73 |
| | 6¢ | $109 |
| | 8¢ | $146 |
| | 10¢ | $182 |
| | 12¢ | $218 |

Ask your salesperson or local utility for the energy rate (cost per kilowatt hour) in your area.

**Important** Removal of this label before consumer purchase is a violation of federal law (42 U S C 6302)

**Figure B**

Figure B shows a sample **energy guide** label. This one indicates the estimated yearly cost to run a certain refrigerator.

1. The estimated cost to run this refrigerator for one year is

   _____$124.00_____ in a region that

   charges _____6.75¢_____ per kilowatt hour.

2. How much would it cost to run this refrigerator for a year in these kilowatt/ hour areas?

   **a)** 12¢ _____$218.00_____     **d)** 2¢ _____$36.00_____

   **b)** 6¢ _____$109.00_____     **e)** 8¢ _____$146.00_____

   **c)** 10¢ _____$182.00_____

3. Does energy cost the same throughout the country? _____No_____

   How do you know? _Different costs per kilowatt hour in different parts of the_

   _country are determined by the local costs of generating electricity._

4. The government tested many similar refrigerators to arrive at its figure of $124.00 to operate this appliance for a year.

   What was the <u>lowest</u> cost estimated? _____$92.00_____

   What was the <u>highest</u> cost? _____$179.00_____

5. Compared to the $124.00 operating cost for this model,

   the <u>highest</u> operating-cost refrigerator would cost _____$55.00_____

   _____more_____ to run.
   <sub>more, less</sub>

6. The <u>lowest</u> operating-cost refrigerator would cost $_____32.00_____

   _____less_____ to run.
   <sub>more, less</sub>

7. Look at the label. On average, the operating cost of this refrigerator was

   _____less_____ than the other models tested.
   <sub>less, more</sub>

8. What, on the label tells us this? _The black bar below the estimated operating cost_

   _compares this model to other models tested._

## FILL IN THE BLANK

*Complete each statement using a term or terms from the list below. Write your answers in the spaces provided. Some words may be used more than once.*

| | | |
|---|---|---|
| higher | filter | longer |
| lower | night | heating |
| daytime | clean | window |
| warmly | high | five |
| less | hot | reducing |

1. Dress _____warmly_____ at home when it is cold. Don't rely on house

   _____heating_____ .

2. Keep the _____filter_____ in your air conditioner _____clean_____ .

3. In cold weather, don't set the house heat too _____high_____. During the

   daytime, keep the temperature no _____higher_____ than 70°F. At

   _____night_____, lower the temperature about _____five_____ degrees.

4. _____Lower_____ your heat when you are not at home.

5. Avoid using lights during the _____daytime_____. Use sunlight. Work close to a

   _____window_____ for light.

6. Use dimmers for incandescent bulbs. You will use _____less_____ energy. The

   bulbs will last _____longer_____ too!

7. Get a flow- _____reducing_____ shower head.

8. Set your _____hot_____ water heater's temperature no _____higher_____ than
   140 degrees F.

1. Make a list of the ways you can conserve electricity at home. Do this for each room at home. Start with the kitchen.

   Kitchen

   _____

   _____

   _____

   Living Room

   _____

   _____

   Bedroom

   _____

   _____

   Bathroom

   _____

   _____

2. Electricity is not the only area of energy conservation.

   Much energy can be saved by following certain driving and car-care rules. You have been around cars long enough to know some of these rules. List them. Ask your

   family members to join in. Let this exercise be a "family affair." _____

   _____

   _____

**CLOTHING PROTECTION** • A lab coat protects clothing from stains. • Always confine loose clothing.

 **EYE SAFETY** • Always wear safety goggles. • If anything gets in your eyes, flush them with plenty of water. • Be sure you know how to use the emergency wash system in the laboratory.

 **FIRE SAFETY** • Never get closer to an open flame than is necessary. • Never reach across an open flame. • Confine loose clothing. • Tie back loose hair. • Know the location of the fire-extinguisher and fire blanket. • Turn off gas valves when not in use. • Use proper procedures when lighting any burner.

 **POISON** • Never touch, taste, or smell any unknown substance. Wait for your teacher's instruction.

 **CAUSTIC SUBSTANCES** • Some chemicals can irritate and burn the skin. If a chemical spills on your skin, flush it with plenty of water. Notify your teacher without delay.

 **HEATING SAFETY** • Handle hot objects with tongs or insulated gloves. • Put hot objects on a special lab surface or on a heat-resistant pad; never directly on a desk or table top.

 **SHARP OBJECTS** • Handle sharp objects carefully. Never point a sharp object at yourself, or anyone else. • Cut in the direction away from your body.

**TOXIC VAPORS** • Some vapors (gases) can injure the skin, eyes, and lungs. Never inhale vapors directly. Use your hand to "wave" a small amount of vapor towards your nose.

 **GLASSWARE SAFETY** • Never use broken or chipped glassware. • Never pick up broken glass with your bare hands.

 **CLEAN UP** • Wash your hands thoroughly after any laboratory activity.

 **ELECTRICAL SAFETY** • Never use an electrical appliance near water or on a wet surface. • Do not use wires if the wire covering seems worn. • Never handle electrical equipment with wet hands.

 **DISPOSAL** • Discard all materials properly according to your teacher's directions.

THE METRIC SYSTEM

## METRIC-ENGLISH CONVERSIONS

| | SI to English | English to SI |
|---|---|---|
| Length | 1 kilometer = 0.621 mile (mi)<br>1 meter = 3.28 feet (ft)<br>1 centimeter = 0.394 inch (in) | 1 mi = 1.61 km<br>1 ft = 0.305 m<br>1 in = 2.54 cm |
| Area | 1 square meter = 10.763 square feet<br>1 square centimeter = 0.155 square inch | 1 ft$^2$ = 0.0929 m$^2$<br>1 in$^2$ = 6.452 cm$^2$ |
| Volume | 1 cubic meter = 35.315 cubic feet<br>1 cubic centimeter = 0.0610 cubic inches<br>1 liter = .2642 gallon (gal)<br>1 liter = 1.06 quart (qt) | 1 ft$^3$ = 0.0283 m$^3$<br>1 in$^3$ = 16.39 cm$^3$<br>1 gal = 3.79 L<br>1 qt = 0.94 L |
| Mass | 1 kilogram = 2.205 pound (lb)<br>1 gram = 0.0353 ounce (oz) | 1 lb = 0.4536 kg<br>1 oz = 28.35 g |
| Temperature | Celsius = 5/9 (°F − 32)<br>0°C = 32°F (Freezing point of water)<br>100°C = 212°F<br>(Boiling point of water) | Fahrenheit = 9/5°C + 32<br>72°F = 22°C (Room temperature)<br>98.6°F = 37°C<br>(Human body temperature) |

## METRIC UNITS

The basic unit is printed in capital letters.

| Length | Symbol |
|---|---|
| Kilometer | km |
| METER | m |
| centimeter | cm |
| millimeter | mm |

| Area | Symbol |
|---|---|
| square kilometer | km$^2$ |
| SQUARE METER | m$^2$ |
| square millimeter | mm$^2$ |

| Volume | Symbol |
|---|---|
| CUBIC METER | m$^3$ |
| cubic millimeter | mm$^3$ |
| liter | L |
| millimeter | mL |

| Mass | Symbol |
|---|---|
| KILOGRAM | kg |
| gram | g |

| Temperature | Symbol |
|---|---|
| degree Celsius | °C |

## SOME COMMON METRIC PREFIXES

| Prefix | Meaning |
|---|---|
| micro- | = 0.0000001, or 1/1,000,000 |
| milli- | = 0.001, or 1/1000 |
| centi- | = 0.01, or 1/100 |
| deci- | = 0.1, or 1/10 |
| deka- | = 10 |
| hecto- | = 100 |
| kilo- | = 1000 |
| mega- | = 1,000,000 |

## SOME METRIC RELATIONSHIPS

| Unit | Relationship |
|---|---|
| kilometer | 1 km = 1000 m |
| meter | 1 m = 100 cm |
| centimeter | 1 cm = 10 mm |
| millimeter | 1 mm = 0.1 cm |
| liter | 1 L = 1000 mL |
| milliliter | 1 mL = 0.001 L |
| tonne | 1 t = 1000 kg |
| kilogram | 1 kg = 1000 g |
| gram | 1 g = 1000 mg |
| centigram | 1 cg = 10 mg |
| milligram | 1 mg = 0.001 g |

# GLOSSARY/INDEX

**absorb:** to take in, 24

**alloy:** two or more metals melted together, 158

**alternating current:** electric current that reverses direction of flow, 196

**ampere:** unit for measuring the number of electrons moving past a point in a circuit, 152

**angle of incidence:** the angle between the incident ray and the normal, 64

**angle of reflection:** the angle between the reflected ray and the normal, 64

**atom:** the smallest part of an element that has all of the characteristics of that element, 114

**circuit:** a path that ends at the same point where it starts, 122

**concave lens:** a lens that is curved inward, 90

**cones:** nerve cells that are sensitive to color, 96

**convex lens:** a lens that is curved outward, 90

**decibel:** a unit that measures the loudness of sound, 36

**density:** the mass of a given volume, 70

**direct current:** electric current flowing in one direction, 196

**domain:** a group of lined-up atoms, 172

**echo:** a reflected sound, 24

**electromagnet:** a temporary magnet made by using electrical current, 190

**electromagnetic spectrum:** radiant energy of all frequencies, from radio waves to cosmic rays, 78

**electromotive force:** electrical pressure, 152

**electrons:** negatively charged particles in the atom, 122

**energy:** the ability to make things move, 8

**farsightedness:** blurred vision caused when light rays converge beyond the retina, 102

**filter:** a transparent substance that transmits some colors and absorbs others, 84

**frequency of vibration** how often an object vibrates in one second, 16

**friction:** the rubbing of one thing against another thing, 114

**galvanometer:** a device that measures weak electric current, 209

**generator:** a machine that changes one form of energy into electrical energy, 122, 208

**geothermal:** relating to heat produced within the earth, 218

**hertz:** a unit that measures the frequency of vibration, 17

**illuminated object:** an object that light shines upon, 54

**image:** visual impression made by reflection or refraction, 90

**incident ray:** a ray of light that strikes an object, 64

**induction coil:** a device that increases the voltage of direct current, 204

**laser:** very strong, concentrated, single-color light, 108

**Law of Reflection:** the angle of incidence is equal to the angle of reflection, 64

**lens** (1): a transparent material that refracts light in a definite way, 90

**lens** (2): a refracting part of the eye that changes shape to focus light rays, 96

**lodestone:** a rock that is a magnet, 158

**loudness:** the amount of energy a sound has, 36

**luminous object:** an object that gives off its own light, 54

**magnet:** a metal that can attract certain other metals, 158

**magnetic field:** the space around a magnet where the force of the magnet is felt, 164

**magnetic force:** the push or pull of a magnet upon a magnetic object, 164

**magnetic induction:** the use of a magnetic field to make a magnetic substance into a magnet, 184

**magnetite:** another name for lodestone, 158

**magnifies:** makes larger, 90

**medium:** a substance through which sound energy moves, 8

**minifies:** makes smaller, 90

**molecules:** very small parts of matter, 2

**natural frequency:** the frequency at which an object vibrates best, 30

**nearsightedness:** blurred vision caused when light rays converge in front of the retina, 102

**neutral:** having no electric charge, 114

**normal:** a line that makes a right angle to a surface, 64

**ohm:** unit of electrical resistance, 152

**opaque:** allowing no light to pass through, 58

**optic nerve:** nerve that connects the eye to the brain, 96

**parallel circuit:** an electrical hook-up in which the current has more than one path, 136

**pitch:** how high or low a sound is, 16

**prism:** a special glass that bends light rays, 78

**ray:** a single beam of light, 64

**reflect:** to bounce off, 24

**reflected ray:** a ray of light that is bounced off an object, 64

**refraction:** the bending of light as it passes at an angle from one medium to another, 70

**resistance:** tendency to slow or stop electric current, 146

**resonance:** the ability of an object to pick up energy waves of its own natural frequency, 30

**retina:** the nerve layer of the eye, 96

**right angle:** a 90° angle, like any corner of a square, 48

**rods:** nerve cells that are sensitive to brightness, 96

**series circuit:** an electrical hook-up in which the current has only one path, 130

**soft iron:** iron that loses its magnetism easily, 178

**sound:** a form of energy caused by vibration, 2

**spectrum:** a band of radiation that includes different frequencies, 78

**static electricity:** electric charges that are not moving, 114

**transformer:** a device that changes the voltage of alternating current, 196

**translucent:** letting light, but no detail, pass through, 58

**transmitted:** passed through, 58

**transparent:** letting light and detail pass through; clear, 58

**transverse wave:** an energy wave that vibrates at right angles to its length, 48

**vacuum:** the absence of matter, 48

**vibrate:** to move back and forth very rapidly, 2

**visible spectrum:** radiation that we can see, including the seven colors of the rainbow, 78

**volt:** unit for measuring electrical pressure, 152